Solving Mechanical Design Problems with Computer Graphics

MECHANICAL ENGINEERING

A Series of Textbooks and Reference Books

EDITORS

L. L. FAULKNER

Department of Mechanical Engineering
The Ohio State University
Columbus, Ohio

S. B. MENKES

Department of Mechanical Engineering
The City College of the
City University of New York
New York, New York

1. Spring Designer's Handbook, *by Harold Carlson*
2. Computer-Aided Graphics and Design, *by Daniel L. Ryan*
3. Lubrication Fundamentals, *by J. George Wills*
4. Solar Engineering for Domestic Buildings, *by William A. Himmelman*
5. Applied Engineering Mechanics: Statics and Dynamics, *by G. Boothroyd and C. Poli*
6. Centrifugal Pump Clinic, *by Igor J. Karassik*
7. Computer-Aided Kinetics for Machine Design, *by Daniel L. Ryan*
8. Plastics Products Design Handbook, Part A: Materials and Components; Part B: Processes and Design for Processes, *edited by Edward Miller*
9. Turbomachinery: Basic Theory and Applications, *by Earl Logan, Jr.*
10. Vibrations of Shells and Plates, *by Werner Soedel*
11. Flat and Corrugated Diaphragm Design Handbook, *by Mario Di Giovanni*
12. Practical Stress Analysis in Engineering Design, *by Alexander Blake*
13. An Introduction to the Design and Behavior of Bolted Joints, *by John H. Bickford*
14. Optimal Engineering Design: Principles and Applications, *by James N. Siddall*
15. Spring Manufacturing Handbook, *by Harold Carlson*
16. Industrial Noise Control: Fundamentals and Applications, *edited by Lewis H. Bell*
17. Gears and Their Vibration: A Basic Approach to Understanding Gear Noise, *by J. Derek Smith*

18. Chains for Power Transmission and Material Handling: Design and Applications Handbook, *by the American Chain Association*

19. Corrosion and Corrosion Protection Handbook, *edited by Philip A. Schweitzer*

20. Gear Drive Systems: Design and Application, *by Peter Lynwander*

21. Controlling In-Plant Airborne Contaminants: Systems Design and Calculations, *by John D. Constance*

22. CAD/CAM Systems Planning and Implementation, *by Charles S. Knox*

23. Probabilistic Engineering Design: Principles and Applications, *by James N. Siddall*

24. Traction Drives: Selection and Application, *by Frederick W. Heilich III and Eugene E. Shube*

25. Finite Element Methods: An Introduction, *by Ronald L. Huston and Chris E. Passerello*

26. Mechanical Fastening of Plastics: An Engineering Handbook, *by Brayton Lincoln, Kenneth J. Gomes, and James F. Braden*

27. Lubrication in Practice, Second Edition, *edited by W. S. Robertson*

28. Principles of Automated Drafting, *by Daniel L. Ryan*

29. Practical Seal Design, *edited by Leonard J. Martini*

30. Engineering Documentation for CAD/CAM Applications, *by Charles S. Knox*

31. Design Dimensioning with Computer Graphics Applications, *by Jerome C. Lange*

32. Mechanism Analysis: Simplified Graphical and Analytical Techniques, *by Lyndon O. Barton*

33. CAD/CAM Systems: Justification, Implementation, Productivity Measurement, *by Edward J. Preston, George W. Crawford, and Mark E. Coticchia*

34. Steam Plant Calculations Manual, *by V. Ganapathy*

35. Design Assurance for Engineers and Managers, *by John A. Burgess*

36. Heat Transfer Fluids and Systems for Process and Energy Applications, *by Jasbir Singh*

37. Potential Flows: Computer Graphic Solutions, *by Robert H. Kirchhoff*

38. Computer-Aided Graphics and Design, Second Edition, *by Daniel L. Ryan*

39. Electronically Controlled Proportional Valves: Selection and Application, *by Michael J. Tonyan, edited by Tobi Goldoftas*

40. Pressure Gauge Handbook, *by AMETEK, U.S. Gauge Division, edited by Philip W. Harland*

41. Fabric Filtration for Combustion Sources: Fundamentals and Basic Technology, *by R. P. Donovan*

42. Design of Mechanical Joints, *by Alexander Blake*

43. CAD/CAM Dictionary, *by Edward J. Preston, George W. Crawford, and Mark E. Coticchia*

44. Machinery Adhesives for Locking, Retaining, and Sealing, *by Girard S. Haviland*

45. Couplings and Joints: Design, Selection, and Application, *by Jon R. Mancuso*

46. Shaft Alignment Handbook, *by John Piotrowski*

47. BASIC Programs for Steam Plant Engineers: Boilers, Combustion, Fluid Flow, and Heat Transfer, *by V. Ganapathy*

48. Solving Mechanical Design Problems with Computer Graphics, *by Jerome C. Lange*

49. Plastics Gearing: Selection and Application, *by Clifford E. Adams*

50. Clutches and Brakes: Design and Selection, *by William C. Orthwein*

51. Transducers in Mechanical and Electronic Design, *by Harry L. Trietley*

OTHER VOLUMES IN PREPARATION

Solving Mechanical Design Problems with Computer Graphics

JEROME C. LANGE

Milwaukee Area Technical College
Milwaukee, Wisconsin

MARCEL DEKKER, INC.   New York and Basel

ISBN 0-8247-7479-5

MARCEL DEKKER, INC.
270 Madison Avenue, New York, New York 10016

Current printing (last digit):
10 9 8 7 6 5 4 3 2 1

PRINTED IN THE UNITED STATES OF AMERICA

Preface

Engineering and design problems that once could be solved only by experimenting with prototypes are now being solved by graphics models generated on interactive computer graphics systems.

The intent of this book is to acquaint the reader with interactive computer graphics and how they are being used in the analysis of mechanical design problems. The book covers four mechanical design topics: the graphics model; mass properties; stress and strain; and kinematic and kinetic analysis. Each chapter begins with a brief review of a particular mechanical design topic, followed by a complete explanation of how interactive computer graphics can solve the problem. This approach enables the reader to compare solutions obtained by conventional methods to those obtained using interactive computer graphics.

The Computervision Design V Computer Graphics System with CADDS 4 software is used throughout the book, except in Chapter 5, where generic software is used to analyze problems in the area of kinematics and kinetics.

Computer-Automated/Assisted/Augmented/Aided Engineering (CAE) is becoming a widely used tool to help build the products of the 1980s. The application of CAE enables designers and engineers to develop better, more reliable products in less time than conventional prototyping methods would take. It is my hope that this book will benefit those who are interested in this subject.

I would like to thank Computervision Corporation and Ken Dabbs for assistance in the preparation of this book. Also, I would like to express my appreciation to Judy for typesetting most of the labels for the illustrations in the book.

Jerome C. Lange

Contents

Preface   iii

1 Introduction   1

1.1	The Meaning of Mechanical Design	1
1.2	What Are Interactive Computer Graphics?	2
1.3	Typical Components of an Interactive Computer Graphics System	6
1.4	Typical Interactive Computer Graphics System (Computervision)	14

2 Creating the Graphics Model   25

2.1	Introduction	25
2.2	Part Structure, Command Syntax, Log In/Out, Part Setup	26
2.3	Creating and Editing Entities	36
2.4	Line Fonts, Centerlines, Crosshatching, and Grids	48
2.5	Display Control: Zoom and Scroll Drawing	60
2.6	Graphics Manipulation: Mirror, Translate, and Rotate Entities	63
2.7	Layers	67
2.8	Dynamic Movement	69
2.9	Dimensioning and Tolerancing	72
2.10	The Three-Dimensional Model: Part Depth, Views, and Construction Planes	80
2.11	Working with Views: The Three-Dimensional Model	98
2.12	Surfaces	116
2.13	Surface Graphics	128
2.14	Cutting Planes and Surfaces	134

v

3   Mass Properties   143

    3.1   Basic Concepts   143
    3.2   Mass Property Calculations (MPC) Using an ICGS   166
    3.3   Mass Properties Calculation Commands   174

4   Stress and Strain   221

    4.1   Stress   221
    4.2   Strain   221
    4.3   Hooke's Law   224
    4.4   Uniform Axial Stress   226
    4.5   Axial Strain and Deformation   227
    4.6   Uniform Shear Stress   228
    4.7   Warping of Cross Section   228
    4.8   Special Case of a Circular Cross Section   230
    4.9   Shear Flow: Shear Stress in Thin-Walled Closed
          Sections   232
    4.10   Flexural Stress in Straight Bars   234
    4.11   Elastic Curve of a Beam   236
    4.12   Relations Between Load, Shear, and Moment   237
    4.13   Shear Stress in a Beam of Any Cross Section   239
    4.14   Axial Stress Superposed on Flexure   241
    4.15   Flexure Superposed on Flexure: General Bending   243
    4.16   Torsion Superposed on Flexure   245
    4.17   State of Stress at a Point   247
    4.18   State of Strain at a Point   256
    4.19   The Finite Element Method   258
    4.20   Finite Element Modeling Using Interactive Computer
          Graphics   264
    4.21   Finite Element Modeling Commands   265
    4.22   Nodes (Grid Points)   266
    4.23   Elements   273
    4.24   Material Descriptions   288
    4.25   Element Property Descriptions   289
    4.26   Loads   294
    4.27   Constraints   295
    4.28   Preparing FEM Data for Finite Element Analysis   297
    4.29   Graphic Display of FEA Results   299
    4.30   Finite Element Analysis   299

5   Design of Mechanisms and Machinery   315

    5.1   Introduction   315
    5.2   Displacement of Mechanisms   315
    5.3   Displacement, Velocity, and Acceleration   321

5.4	Instant Centers	323
5.5	Velocity and Acceleration in Plane Motion	325
5.6	Coriolis Acceleration	329
5.7	Force Analysis of Mechanisms	331
5.8	Motion Analysis Using Interactive Computer Graphics	346
5.9	Velocity Analysis Using Interactive Computer Graphics	358
5.10	Acceleration Analysis Using Interactive Computer Graphics	366
5.11	Kinetic (Force) Analysis Using Interactive Computer Graphics	370

Bibliography 393

Index 395

Solving Mechanical
Design Problems
with Computer Graphics

1
Introduction

1.1 THE MEANING OF MECHANICAL DESIGN

Engineering is one of the most important of all human activities. The pyramids of Egypt and the aqueducts of ancient Rome were early engineering accomplishments. Engineering produced the steam engine and the machines it powered, which allowed the average person to have some of the things that had previously been available only to the rich. It was engineering that achieved the airplane, fulfilling one of man's oldest dreams. It is engineering that has made it possible for us to go to the moon. Engineering has the responsibility of providing us with those things that make our lives easier and more interesting.

Design is a creative activity, and it is not an exaggeration to say that our creative efforts have produced those things that most distinguish us from other forms of life. Creativity as it applies to engineering is the ability to conceive basic innovations, to perceive those problems that can readily be solved, to devise solutions to new problems, and to combine familiar concepts in unusual ways. Thus design is the creative part of engineering. The word *design* will hereafter be used for this aspect of engineering and *designer* for a person who engages in this activity.

The end result of the design process will hereafter be called the *product*. Many features desired in a product are incompatible; thus it is readily apparent that design involves making decisions and comprises based on careful consideration of many factors. It also becomes apparent that the designer must continuously exercise judgment and that faulty judgment in only one instance could invalidate an entire design concept.

Design is a large field and it is becoming larger very rapidly as discoveries are made, as inventions are produced, and as new needs

arise. These inventions or discoveries do not benefit us directly;
it is only when they serve as the basis for a design or are incorporated
into a design that we derive a benefit from them. It is also appropriate
to point out that a new design is itself of no real significance unless
it clearly fulfills a need and is produced in an appropriate quantity.

A designer is concerned with such problems as the geometric ar-
rangement of components, the effect that the motion of one part has
on the motions of those associated with it, and the effects of forces.
Designers are also concerned with the properties of materials, with
the capabilities and limitations of manufacturing processes, with human
capabilities and limitations, and with economic matters. Designers
must also possess a highly developed ability to present their ideas
concerning complex engineering problems in such a manner that they
are readily and clearly understood by other technical personnel.

Mechanical design is the application of many of the principles of
science and technology in the creation of a product and the consideration of the various factors that affect its production and use.

An understanding of the relation of drafting to design is important
to an appreciation of the position of design. Drafting is the preparation
of drawings that serve as the means by which the engineering department exactly describes the various parts and assemblies that make
up a product, and this requires some ability in graphic science. A
designer also makes considerable use of graphic science, which is most
conveniently done at a drafting table and/or through a computer graphics system. But to say that the work of designers is drafting would
be the same as saying that the work of an author who uses a typewriter is merely typing.

1.2 WHAT ARE INTERACTIVE COMPUTER
GRAPHICS?

A new era of enhanced creativity and productivity is possible with
the application of computer graphics to the areas of mechanical drafting, design, and engineering. Interactive computer graphics is a part-
nership between the computer and the human being, combining the
best abilities of each to form a powerful drafting, design, and engineering team. This interaction between people and computers is accomplished through the use of cathode-ray-tube (CRT) displays, alphanumeric keyboards, digitizing tablets and pens, and special computer
languages.

A person is able to communicate or interact with the computer and
receive a direct response from it. This two-way conversation may
be graphical or pictorial in nature, with the response time being only
seconds. For example, the designer/engineer may generate a picture
on the cathode-ray display tube by using a digitizing table and a pen.

Due to previous programming, the computer understands the picture, makes calculations based on it, and presents answers or a revised picture to the designer/engineer within a few seconds. The process can be repeated, simulating the same process that is performed on a conventional drafting board, but at a faster rate.

Most interactive computer graphics systems (ICGS) provide the same capabilities as the drawing instruments used by the drafter, designer, or engineer, so that the user can construct a drawing on a CRT display in a manner similar to that used to draw on a drafting board.

A part is drawn on a ICGS by using a graphical vocabulary for constructing points, lines, circles, arcs, and so on. The operator can call up the graphical vocabulary by use of a menu or by means of a pushbutton panel which enables the operator to use the following graphical constructions. The following graphical constructions represent only a small portion of the constructions that are available on most interactive computer graphics systems.

A point may be placed by using any of these options:

1. Locating the point by means of a digitizing pen and a tracking cross
2. Inputting the coordinates via the keyboard
3. Designating the intersection of lines and/or arcs

A line may be drawn by using any of these options:

1. Drawing it between two existing points
2. Locating the line's endpoints via the digitizing pen
3. Inputting the coordinates of the line's endpoints via the keyboard
4. Drawing it through a point, tangent to a circle
5. Drawing it parallel to an existing line at a specified distance from it
6. Drawing it through a point at a specified angle to an existing line
7. Drawing it tangent to two circles

A circle may be drawn by using any of these options:

1. Drawing it with a specified center and radius
2. Drawing it with a specified center, tangent to a line
3. Drawing it with a specified center, tangent to another circle

A circular arc may be drawn by using any of these options:

1. Drawing it tangent to two lines, with a specified radius
2. Drawing it tangent to a line and an arc, with a specified radius
3. Drawing it tangent to two arcs, with a specified radius

In these constructions, the digitizing pen is used to select points, lines, and circles as required. In addition to being able to use the basic construction features, operators can change the scale and center of the picture at will, establish whatever local coordinate systems they wish, and erase selected elements.

Interactive computer graphics enables the designer/engineer to draw a three-dimensional mathematical model of an object in space. All points. lines, and planes making up this model are defined mathematically in space. Therefore, the designer/engineer has a powerful tool available for calculating distances, angles, lengths, areas, volumes, and so on, of geometric elements in space with extreme quickness and accuracy. Points, lines, and planes making up the object can be manipulated (rotated, moved, copied, stretched, etc.) and stored in files with ease using interactive computer graphics. If the design needs to be altered or improved for any reason, the time for carrying out this redesign is minimal on a interactive computer graphics system. The more design cycles that can be carried out within the limitations of time or budget, the better the results.

The benefits of a interactive computer graphics system can be summarized as follows:

1. The user can quickly see and correct any mistakes on a drawing.
2. A three-dimensional view can be constructed and shown on the ICG's CRT display. This three-dimensional view is a spatial model and is beneficial to the designer or engineer during the design process and for engineering analysis work.
3. Repetitive tasks can be preprogrammed as part of the computer's software. The computer performs the monotonous tasks, so that the operator can spend more time in creative design and problem solving.
4. The data generated by a interactive computer graphics system can be used easily by other computers. This feature is very important when it comes to analysis, manufacturing, robotics, forecasting, marketing, scheduling, and so on.

Attempting to explain how the human element works, in the same terms as those used for electromechanical devices, is a challenge to medical science. Identifying what the human being does in an ICGS is a little easier. First, the person has an idea of how to solve the geometry-related problem before beginning to work on the ICGS. These ideas can be in the form of sketches, drawings, and/or in conceptual form in one's mind. This idea must be entered in the ICGS as a three-dimensional object as it would appear in the real world. This object, known as a three-dimensional graphics model, is defined mathematically in space by points, lines, and planes.

On this graphics model, one can perform various problem-solving solutions, depending on one's needs. For example, one could generate information concerning how the object could be manufactured or how it would function under actual working conditions. Therefore, the three-dimensional graphics model becomes the explicit data base, comparable to an engineering drawing.

This book is concerned only with the analysis of an object as related to mechanical design. Figure 1.1 illustrates what happens after the graphics model is generated on an ICGS. This process will be referred to as the *interactive computer graphics analysis process.*

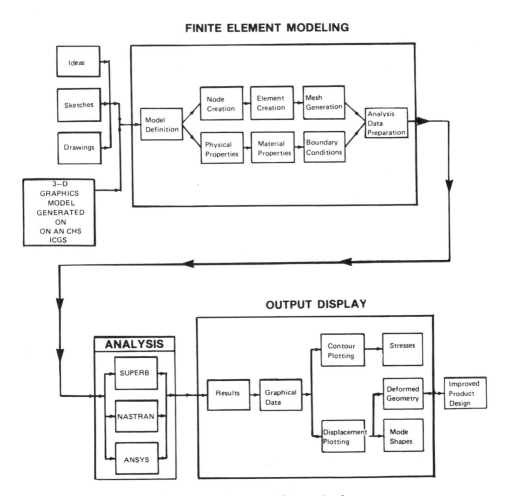

FIGURE 1.1 Interactive computer graphics analysis process.

The steps that are followed in the mechanical design analysis of a three-dimensional graphics model (object) on an ICGS are as follows:

1. The model is generated on an ICGS.
2. The model is prepared for finite element modeling (commonly known as *postprocessing*).
3. After the model is prepared (model definition), the various nodes, elements, and mesh generation are inputted graphically using the ICGS. This results in a finite element model of the graphics model. Also, the various physical and material properties and boundary conditions are defined for the finite element model during this step.
4. The finite element model is prepared (postprocessing) for various analysis programs, such as SUPERB, NASTRAN, or ANSYS.
5. The prepared finite element model is submitted to the analysis program, which is usually run on a computer other than the ICGS's computer, because of the large amount of computing time and the number of capabilities that are required.
6. After this task has been completed, the results are returned to the ICGS as graphical data. These graphical data can be displayed on the CRT as stresses, deformed geometry (strains), mode shapes, vibrations, or as contour or displacement plots on paper.

This interactive computer graphics analysis process improves the design of the idea (object) because the user can actually see what would happen to an idea under actual working conditions before everything is manufactured. This process is much easier and faster than the methods that have been used in the past.

1.3 TYPICAL COMPONENTS OF AN INTERACTIVE COMPUTER GRAPHICS SYSTEM

All systems in use today consist of a computer, a magnetic tape unit, a magnetic disk unit, a display terminal, a keyboard, an electromagnetic tablet and stylus or light pen, usually a hard-copy device, a large digitizer, a plotter, a medium- to high-speed printer, and software. In terms of their functions these components are usually classified as input, output, display, computational, or storage devices. Some devices, such as tape and disk units, have both input and output capabilities. A typical ICGS is presented schematically in Fig. 1.2.

The Computer

In an ICGS the computer does most of the work. It processes data and stores and manages them. In the process of designing or engineering simple to very complex devices and machines, large amounts of geometric data must be produced and manipulated. Because the com-

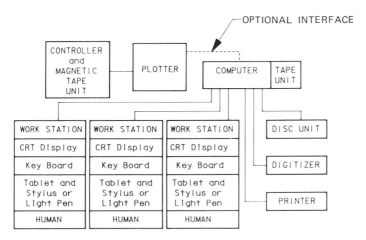

FIGURE 1.2 A typical interactive computer graphics system.

puter possesses great accuracy and incredible speed, has an outstanding memory, and works automatically until a task is completed, it is ideally suited for extending the human brain in the creative process of developing geometric and numerical data.

The computing performed in an ICGS takes place in the central processing unit (CPU) of the system's primary computer. In many systems today this processor is supported by microprocessors programmed to specialize in individual tasks, such as rotating the geometry displayed on one of the terminals. This support helps to speed up the system's response to each user by freeing the CPU to execute tasks for other users. Quite often these microprocessors are located in the display terminals and can perform several tasks, so that the main computer is called on only for large, complex calculations.

The computer's primary memory stores all the instructions and facts required to process each user's task. Its memory capability can be greatly increased by the use of magnetic tape and disk storage devices, which are also used to store information about tasks that are completed and tasks that are awaiting completion.

The Tape Unit

The tape unit is a device that can be classifed as input/output, in that data can be stored on tape and retrieved from it. Two types, magnetic and punch tapes, are used, but magnetic tape is more prevalent. Magnetic tapes have high data storage density, rapid rates of storage and retrieval, and are reusable. They are used most frequently to back up data stored on disk, to transfer data to other machines, and for archival storage. (In the case of archival storage, a system for periodic tape verification plus backup tapes must be used, since data can deteriorate on magnetic tape.) Tapes may come packaged on cassettes, reels, or cartridges. Mylar or paper punch tapes are still used to transfer small amounts of data from one machine to another. The most common use of punch tapes is to provide the data required by numerically controlled manufacturing machines and some older digital plotters.

The Disk Unit

The magnetic disk can also be classified as an input/output device. It is usually called a mass storage device because of the large amounts of data that can be stored. The disk has very high data access and transfer rates and is used to extend the physical storage capability of the computer. Disk units have a wide range of characteristics and sizes. Some units have a fixed disk; the actual disk platter cannot readily be removed. Some units use floppy disks that are removable; they are thin, flexible, and reasonably priced. Other units use removable disk packs ranging from one to six to eight disks. Disks vary in size from 5 1/2 in. up.

The Printer

A printer is a device for listing the alphanumerics going into or out of a computer. Printers use impact, electrostatic, and thermal technologies to place the character on the paper- or plastic-based output material. The actual character generation may be in dot matrix form (with the characters formed by a series of dots) or in fixed form (like a typewriter). The speed and quality of printing varies from about 10 characters per second to over 30,000 lines per minute. The quality ranges from faint to bold and from boxy and hard to read to finely formed and easy to read.

The Workstation

The workstation is the most visible part of an ICGS. It is the interface between the human and machine elements of the system. The workstation has two principal functions: to send messages and data to the computer and to receive the computer's answer. An ICGS presents these answers very rapidly as pictures and numbers that can be assembled or altered with simple commands. This conversation is accomplished using keyboards, electronic tablets, electronic styluses, special-function boards and light pens for sending messages, and some type of CRT for displaying and accessing geometry.

The Keyboard

The keyboard, in most cases, looks and feels like the keyboard of an electric typewriter. It is connected to the computer and is used to send information, to "initialize" the system to start an activity, and to add additional program capabilities when required.

The Tablet and Stylus, Special-Function Keyboard, and Light Pen

The electronic tablet is used in conjunction with the stylus to control the cursor display on the screen so that specific elements can be pointed to and selected. The tablet is also designed so that it can be divided into a matrix of squares. Each square can be assigned a specific function so that, when touched, it causes a certain operation in the computer to take place. The stylus and tablet work together to send preprogrammed messages to the computer. Together they comprise the primary method used to construct the geometry being displayed on the CRT. These components simulate pencil and paper and are both the most powerful and the easiest elements for human beings to use.

Some systems have a unique feature called a special-function keyboard. This device is similar to and supplements the electronic tablet. It is called "special function" because its keys can be programmed and changed rapidly to perform highly repetitive functions peculiar to each task. Whereas the stylus and tablet control functions general to all tasks, the special-function keys allow task-related control. For example, if one were creating an office interior design, special-function keys could be programmed to place different pieces of office furniture in one location on the display; then the stylus and tablet would be used to select and move the furniture to specific locations and orientations.

Light pens are used in some systems in place of the stylus-tablet method of selecting features and functions. In such a system part of the display screen contains a "menu" of commands; touching the screen with the light pen causes the computer to act on the appropriate command. The pen also controls the cursor, which indicates where a given action will take place on the CRT screen, and thus can select certain elements of the geometry for action. An example is the deleting of a line. The work "delete" would be displayed in the menu area of the screen, together with other commands, such as "add," "move," "rotate," "execute," and "digitize." The user would select the word "delete," point to the line, and touch the light pen to the word "execute"; the line then would be deleted.

The CRT Display

The CRT screen is the most visibly active part of the system because of the ever-changing geometry being displayed on it. This portion of the system is truly interactive because it sends, through the eyes to the human brain, the signal that prompts the next step in the creative or construction process.

Interactive graphic systems today use three main types of graphic display devices: storage tube terminals, raster-scan terminals, and vector-refresh terminals. All three types use a phosphor-coated tube and a finely focused electron beam that excites the phospher coating and produces a trace wherever it strikes. Depending on the type of tube used, the image may appear in one of several monochromatic colors, such as black, white, green, or blue. Some tubes display a mix of several colors; others display a full range of colors much like that of a color television. Each type of tube has other distinctive characteristics affecting, for example, resolution, which contributes to picture quality, animation, which provides different motion capabilities; picture brightness; and cost. Each offers some advantages for specific applications.

Storage tubes are used most widely because they require the least amount of computer resource to keep the picture on the screen. Once generated, the image can be stored on the tube, and regeneration by the computer is not necessary to maintain the picture. Storage tubes also have excellent resolution and very fine lines. They do not have color, however, and they require a low-light-level environment for good viewing because of their limited contrast capabilities. The greatest drawback to storage tubes is the lack of dynamic manipulation capability. Images cannot be selectively altered, rotated, or moved without repainting the entire display. This causes a noticeable time lag between command and display.

Vector-refresh terminals are very popular in systems that require animation and continuously changing geometry because they are capable of magnifying, rotating, or translating geometry while the user watches. They can do this because the image is generated by deflecting the electron beam trace directly on the screen in a continuous sweep. Once it is completely defined, the image is refreshed by retracing before the phosphor fades. This repainting is accomplished by storing the image in memory and constantly repainting under computer control. These units have excellent brightness and by creating lines from start point to endpoint (not using a dot matrix) provide very good resolution. Since lines are created individually, selective erasing is possible. Disadvantages of these terminals are limited color and image flicker resulting from large amounts of data (measured in vector inches). Additional hardware and memory are required for image storage, which makes these terminals very expensive.

Raster-scan terminals are excellent for solid modeling, where surfaces can be defined using different colors, and for finite element analysis, to show stress distribution. They usually cost less than other terminals and have excellent brightness and good contrast for viewing in normal room lighting. Their major disadvantages are poor line quality and the high demands they make on computer memory and processing. These terminals use a matrix of dots called pixels to produce a picture. In this system the electron beam intensity continuously changes as the beam sweeps across the screen to activate certain dots. This intensity modulation is controlled within the limits of an intensity scale, and the picture is controlled by a bit map in the computer that corresponds to each dot or pixel on the screen. To maintain the picture without noticeable fade, the screen is scanned about 60 times a second.

The Plotter

In Fig. 1.2 the plotter is shown connected to the computer by a dashed line. The reason for this is that plotters can be linked to and controlled directly by the computer or, as illustrated, can have an independent controller and input device. The most popular input device is a magnetic tape unit. The plot data are prepared on a computer, postprocessed into a form acceptable by the plotter, and written on a magnetic tape. (Postprocessing is a special activity required to convert data in one form to some other form.) The magnetic tape is then mounted on the plotter's tape unit, which reads the data and sends them to the controller. The controller sends out signals to move the drawing head in x and y directions, turns the line-producing signal on or off as required, and controls drawing speed.

Plotters are sometimes categorized as "hard-copy output" devices, a category that includes printers and printer/plotters. Plotters produce lines and characters by a variety of techniques on different materials and are usually identified by the drawing techniques used: pen plotters, photographic plotters, CRT hard-copy units, and electrostatic, laser ink jet, and impact printer/plotters.

Pen plotters are quite versatile because they can draw with ballpoint, technical, and felt-tip pens. Depending on the plotter, up to six different colors can be used. Some pen plotters can be equipped with steel points, diamond points, swivel knives, and lasers. Pens are used for drawing on paper or plastic film, while cutting points are used for scribing drawings on coated plastic, metal, or glass. Lasers are used for cutting special-purpose material such as stencils. The plot size can range from 8.5 by 11 to 60 by 254 in.

Pen plotters may be flatbed, drum, or sliding tables and may be driven by linear, stepping, friction, or servo motors. The signals may be analog or digital and may use rotary encoders or linear measurement systems to determine location. A flatbed plotter has a gantry that moves along the x axis above the drawing surface, and a moving toolholder mounted on the gantry that moves along the y axis. A drum plotter uses a vacuum and/or sprocketed paper holder and rotates in two directions to achieve the x movement. The toolholder is driven parallel and tangent to the axis of the drum to achieve y movements. A sliding table plotter moves the plotting bed under a fixed toolhead. Pen plotters may be freestanding; that is, they may have a controller and tape drive that allow plotting to be independent of any other power

and signal source. Pen plotters can also be integrated into the ICGS
and use the computer as the main signal source.

Photoplotters are essentially the same as flatbed pen plotters except
that the toolhead is a light source projected through lenses, shutters,
and apertures onto photographic film. The most distinctive aspects of
photoplotters are their accuracy, resolution, and line quality. (Accuracy is defined as the amount of tolerance by which an output point
can deviate from the expected absolute location. Resolution is the
smallest increment of measurement that the system is capable of detect-
ing.) Photoplotters usually do not produce drawings larger than the 30
by 40 in. standard film size. Also available but seldom used or needed
is 36-in. roll stock.

CRT hard-copy units are used to reproduce what is displayed on
the screen of the workstation. These units use a band-type CRT located in the hard-copy unit to transfer the image onto silver oxide-
treated paper and heat it to develop and fix the image. At present,
hard-copy output is limited to 8.5 by 11 in. The quality of the image
is limited because fine detail is often blurred or washed out. The
chief advantage of these units is speed.

Electrostatic printer/plotters produce the image on electrosensitive
paper or plastic film by passing the material over a row of nibs that
are in an electrically charged or uncharged state. These nibs produce
small, charged dot areas on the material. As the material passes
through a toner bath, the toner adheres to the charged areas. The
density varies from 100 to 200 dots per inch, and the dots vary in size
and shape depending on the plotter. These units are very fast, produce good drawings, and are usually connected directly to the computer.

Laser plotters produce images on light-sensitive material very much
as the electrostatic printer/plotter does, except that a single laser
beam is used. These units are extremely fast and operate most efficiently if the controller is connected by a high-speed data line.

Ink jet printer/plotters are well described by their name. The
material is usually passed over a roller and under nozzles aimed at the
roller axis. The nozzles, which vary in number depending on the color
tone capabilities of the plotter, squirt colored ink onto the material.

Impact printer/plotters also operate efficiently when linked directly
to a computer. They produce color and black-and-white drawings on
standard computer printing paper. All systems use an inked ribbon
and print hammers that print out the image in dots. The resolution
quality is near the low end of the performance scale because the units
use only 60 to 100 dots per inch. Special controllers or standard computer data links are used to drive these printer/plotters.

The Digitizer

The digitizer is an input device that obtains digital data from drawings. These units have embedded grids in the surface, which resembles

FIGURE 1.3 Computervision's Designer V Graphics System.

a wire mesh, and a cross-hair cursor with button switches or a stylus that can be moved over the surface. When a button is pushed or the stylus is pushed down, a signal is recorded from the nearest grid point of the mesh. This signal represents a coordinate location with respect to the selected starting point and is recorded in the computer. Using a menu of nouns to define what is between the points, lines, arcs, and so on, the drawing geometry can be re-created in the computer for further use. These units have limited use in mechanical application because the intended accuracy varies with the accuracy of the drawing, the accuracy of the operator, and the fineness of the grid.

Software

The term "software" is used to describe all the computer programs, documentation, and operational procedures required to operate a system. Because of the nature of software, it is now shown in the schematic in Fig. 1.2.

1.4 TYPICAL INTERACTIVE COMPUTER
GRAPHICS SYSTEM (COMPUTERVISION)

This section introduces the reader to the Designer V System hardware
(Fig. 1.3). The reader will learn the functions of the two major Designer V hardware components: the control station and the workstation (Fig. 1.4).

FIGURE 1.4 Computervision's Designer V Graphics System.

1.4.1 Designer V System Hardware

Figure 1.5 shows the major components of the Designer V System. The central hardware unit of the system is the Computervision graphics processor (CGP), which is the actual system computer. The CGP is connected to the tape drive unit, disk drive, and the Instaview workstations.

The workstation includes a variety of hardware units (Fig. 1.6). Each plays an integral part in the effective use of the Designer V System.

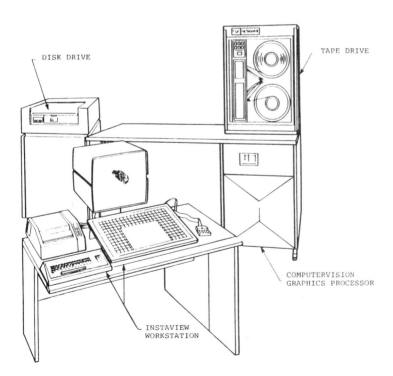

FIGURE 1.5 The Designer V System.

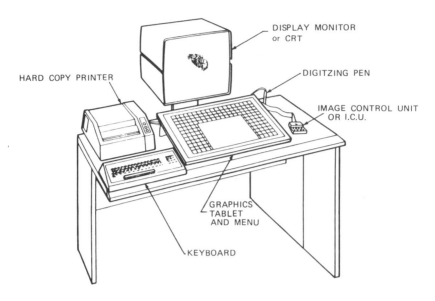

DISPLAY MONITOR
or CRT

HARD COPY PRINTER

DIGITZING PEN

IMAGE CONTROL UNIT
OR I.C.U.

GRAPHICS
TABLET
AND MENU

KEYBOARD

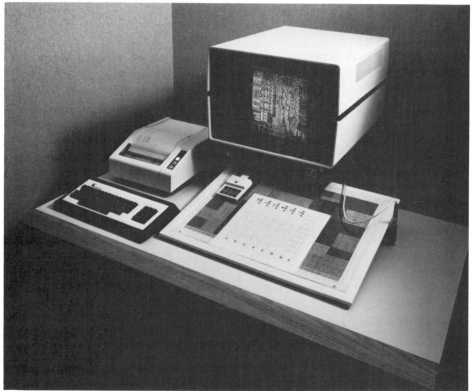

FIGURE 1.6 Instaview work station.

The user will be using the keyboard to communicate with the system. The graphics tablet and menu contain a specified area for "digitizing" locations in part. They also include a menu area that the user will use to input commands and related information by digitizing with the pen rather than typing on the keyboard.

The pen is used to identify locations on the graphics or menu area. The image control unit (ICU) provides a variety of functions which are helpful in the control of the image on the CRT. The CRT, also called a display monitor, is the screen that displays the visual image. Finally, the hard-copy printer will provide the user with a paper copy of the text or the graphics appearing on the screen.

The keyboard has 68 keys which are arranged much like the keys on a typewriter (Fig. 1.7). It also has a few special keys and a second set of number keys on a numeric pad on the right. The numeric pad lets the user enter numbers rapidly.

Many of the keys, such as the space bar and SHIFT keys, work much like typewriter keys. The RETURN is also similar to the return on a typewriter. Its major use, however, is to execute a command.

A single key can function in three ways: (1) when the user presses the key alone, (2) when the user presses the key together with the SHIFT key in order to specify the character at the top of the key, and (3) when the user presses the key and the CONTROL key together to specify the internal function of the key.

A graphics tablet is a drawing table on which the user digitizes using an electronic pen (Fig. 1.8). The user immediately sees the graphics on the graphics display screen.

A menu is a device the user can use to input commands to the system quickly. The menu has squares or pads defined on a digitizing surface (usually found on the graphics tablet). These squares are activated by digitizing them with the electronic pen. The user can define each square so that it activates a command or a series of commands.

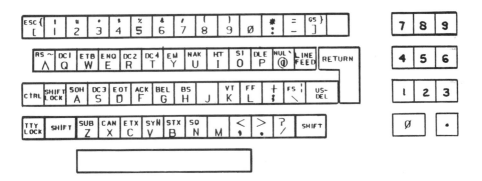

FIGURE 1.7 Alphanumeric keyboard.

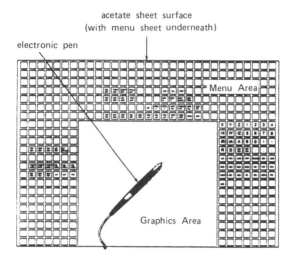

FIGURE 1.8 Graphics tablet, pen, and menu.

This saves time by eliminating the need for the user to type the command manually. Menus are interchangeable; new menu sheets can be placed under the acetate surface of the tablet.

The electronic pen is used to locate positions on the screen that the user wants to digitize. It controls the movement of the cursor, a cross-hair indicator, around the graphics area. Place the point of the pen on the tablet surface, move it to the desired point, and push the bottom button. To select menu keys, aim for the middle of the square.

The graphics display screen or CRT is a display device that allows the user to see graphics such as designs, drawings, and tool paths being created or modified. The CRT also displays the text of the commands that the user has entered. The screen (both text and graphics) automatically goes blank if the user does not press a key within 5 to 6 minutes. The image that was on the display will reappear automatically when the user again presses any key. Commands the user enters through the keyboard and menu are referred to as communication text and appear on the screen (Fig. 1.11).

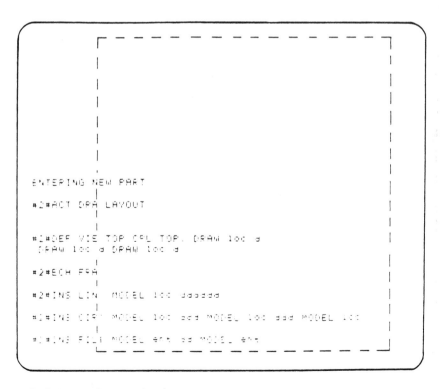

FIGURE 1.9 Communication text on the CRT.

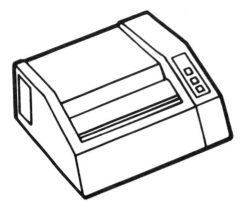

FIGURE 1.10 Printer.

A small table-side printer (Fig. 1.10) provides the user with hard
copies. The printer can copy both text and graphics. These copies
of the graphics are exactly like the images displayed on the CRT. The
printer is not meant to be used for final drawings but rather for quick,
temporary, working copies. It also gives the user a permanent record
of all the commands the user used, thus helping the user document
his or her work. "Control station" is a general term that refers to the
computer and its associated specialized hardware.

A computer is a machine that carries out the necessary processes,
calculations, and operations of a CAD/CAM graphics system. Once the
user enters a command at the workstation, it is sent to the computer,
which decodes and executes it. The computer then processes the information and sends it to the CRT (Fig. 1.9) where the user sees it
displayed. CAD/CAM systems can have several computers within their
total setup, or configuration, to expand the overall performance of the
system.

The CGP 200 (Fig. 1.12) distributes the processing of information
to several processors instead of one. The graphic processing unit
(GPU), the central processing unit (CPU), and the floating-point unit
(FPU) run at the same time in parallel, each having special tasks to
perform. This distributed approach is like having several helpers do
a job instead of just one.

A disk storage device (Fig. 1.13) stores and accesses all the information that is essential to the operation of the system. Information is
stored on a disk or a disk pack. This information can include parts,
consisting of model geometry, detail drawings, and associated manufacturing information.

The magnetic tape unit is a device (Fig. 1.14) that allows the user
to transfer part designs (model geometry), drawings, and manufactur-

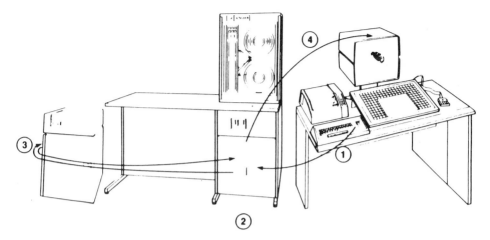

FIGURE 1.11 Command input-to-display sequence: (1) command is entered at workstation and goes to computer; (2) computer decodes and
executes command; (3) information and code form put in temporary
file area on disk; (4) computer sends image to CRT.

ing information from the disk to a magnetic tape. The user can restore
the information an any disk anytime the user wants to use it again.
Magnetic tapes can also be used to transfer software programs to the
system.

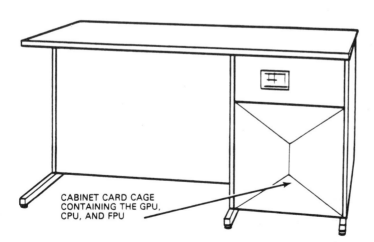

CABINET CARD CAGE
CONTAINING THE GPU,
CPU, AND FPU

FIGURE 1.12 Computervision graphics processor (CGP).

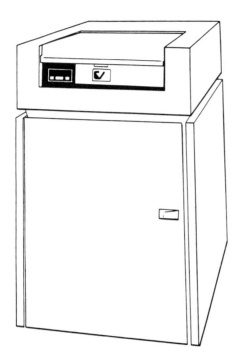

FIGURE 1.13 Disk storage device.

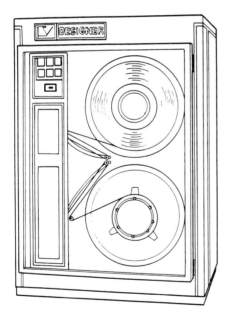

FIGURE 1.14 Magnetic tape unit.

1.4.2 Software

Hardware alone, no matter how sophisticated, cannot accomplish any task without computer programs, known as software. Software consists of a set of internal instructions that direct operation of the computer.

The major operational levels available in the CADDS 4 software are the Operating System (OS), called the system level, and the Computervision Automated Design and Drafting System (CADDS), called the graphics level. The system level is a nongraphic level; its activity concerns the internal management of CADDS 4. For example, the user can tape files off the disk onto magnetic tape at this level. All part creation, design, and detailing operations are performed at the CADDS or graphics level through the CADDS 4 graphic language.

Two operational modes exist within CADDS 4: the Model mode and the Draw mode. They distinguish between the model design operation and the detailing/drawing operation (Fig. 2.1).

2
Creating the Graphics Model

2.1 INTRODUCTION

Computer-aided design and engineering systems simplify and expedite the creation of part geometry. In a process similar to making a drawing or creating a physical model, designers construct three-dimensional computer-based models which provide a data base of graphic and nongraphic part information.

As design information is generated, an image of the part takes shape on the display screen. The engineer or designer can easily manipulate and examine this image by rotating it or zooming in, or can create exploded views of disassembled parts. Overall, a wide range of integrated design and display options offers engineers and designers immediate transformation of mechanical and mathematical concepts into clear visual images.

Using these graphic capabilities of an interactive computer graphics system, designers and engineers examine surface curvature, edges, and intersections, as well as check for interferences, incomplete constructions, and other design violations without constructing a physical prototype. Thus many design alternatives are considered in the time previously required to examine just one.

To illustrate how to construct the graphics model on an interactive computer graphics system, the command syntax related to Computervision's CADDS 4 software will be used throughout this chapter.

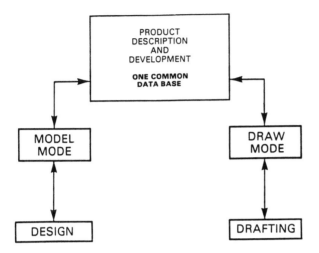

FIGURE 2.1 Multiapplication of a common data base.

2.2 PART STRUCTURE, COMMAND SYNTAX, LOG IN/OUT, PART SETUP

2.2.1 Part Structure: Operation Modes, Model/Draw

The Computervision Automated Drafting and Design System 4 (CADDS 4) has separate modes for the design of a two- or three-dimensional model (Model mode) and for the creation of detailed drawings from the model (Draw mode) (Fig. 2.1). The model geometry created in the Model mode (Fig. 2.2) is accessed by the drafter in the Draw mode,

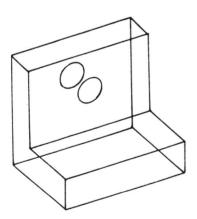

FIGURE 2.2 Model mode.

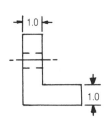

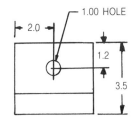

FIGURE 2.3 Draw mode, views of model with dimensions and note.

and two-dimensional drawings are created with dimensions, text, and notes (Fig. 2.3). Any number of drawings can be created in the Draw mode and associated with a specific model. The drawings can be of different sizes and scales. Model entities appear on all drawings. Draw entities appear only on the specific drawing on which they are entered.

2.2.2 Command Syntax

Structure

Instead of mnemonics, CADDS 4 uses normal spelling for most words and establishes abbreviations for compound words. Keywords, however, are shortened to require only the first three or four characters. Required characters are displayed in uppercase and those not required in lowercase: for example,

ACTivate	INSert	CPL = construction plane
PARt	LINe	BSP = B-spline
POInt	HORizontal	SGR = surface graphics

CADDS prompt	Verb/noun phrase	Modifier	Colon	Mode prompt	Location/ entity prompt	Specify loc/ent	Execute command
#n#	INSert LINe	HORizon- tal	:	*Model*	*loc*	$d_1 d_2$	[CR]

Data Input Prompts

These prompts appear automatically depending on the command; they are not typed in by the operator.

Prompt	Type of data requested
Model loc	Location of specification
Model ent	Entity identification
Draw loc	Location of specification
Draw ent	Entity identification
View	View selection
Model/Draw ent	For command applicable to both modes
Model/Draw end	End of entity

. Punctuation

Blank space: required between keywords. INS LIN
: Colon: used to enter data and to execute command and return mode prompt. INS LIN: *Model loc* d_1d_2 : *Model loc* d_3d_4 [CR]
, Comma: used to separate explicit coordinates. *Model loc* X3, Y4, Z6
; Semicolon: used to switch from identifying entities to specified location. *Model ent* d_1; *Model loc* d_2d_3 [CR]
. Period: used as decimal point or to execute command and enter new modifiers. INS LIN HOR : *Model loc* d_1d_2. VER : *Model loc* d_3d_4. PAR : *Model ent* d_5 (etc.)

Location and Specification Masks

END	End of an entity.
ORG	Origin of an entity.
LOC	Location cancels out previous locator.
INTOF	Intersection of two entities.
SET	Sets a coordinate value until command completed.
Nn	Number of times command repeated. n = additional copies.

2.2.3 Function/Control Keys

[CR]	Carriage return. Executes command.
↑Rn	Control R task number. Initiates logging in.
ESC	Escapes. Stops text output.
ESC Q	Quits text output and returns to CADDS.
↑H	Allows backspacing while entering verb or noun and while entering part name and text.
↑E or ↑Q	Aborts command prior to execution.

US-DEL	User-delete key negates data input entry prior to execution.
↑CC	Aborts task—very powerful. Returns system to Operation System level.
↑L	To log in to lowest-available task number.
#n#	CADDS-level prompt.
n>	System-level prompt.

Note: The symbol indicates that the CTRL key is to be held down while the next key is pressed.

2.2.4 Log In File/Log Out

This material covers the procedure for entering and exiting the CADDS graphics system.

Log In

The user will be seated in front of a blank screen and enter the following through the keyboard. In this explanation, the left side indicates what is entered by the user and the right side indicates comments.

↑Rn   (To activate the terminal and enter the system level.)

* * * TASK n INITIATED * * *

n> CADDS [CR]   (At the system level, CADDS is en-
ENTERING CADDS4 REVn   tered to enter the graphics level.)

#n#   (Now at the graphics level and ready to create a part. See Section 2.2.5.)

File

The system provides for filing data without exiting the part with the option of assigning a new part name.

Syntax

Filing under current part name:

#n# FILE Part [CR]   (System remains at the graphics level.)

Filing under new part name:

#n# FILE PART newname [CR] (System remains at the graphics level.)

Log Out

The EXIT PART command is the first step in logging out.

Syntax

#n# EXIT PAR mod [CR]

Modifiers

Q   Quit part and do not file.
F   F, followed by return, files the part under the
                         original part name.
F new part name   Assign new part name by adding the name fol-
                         lowing F.
OS   Return to system level. Used in conjunction
                         with Q or F.

Example 1

a. #n# EXIT PAR Q [CR]   (Part not filed.)
b. #n# EXIT CADDS [CR]   (New/old part name could be ac-
                                 tivated at this time.)
c. n> LOGOUT [Cr]   (Entered to terminate task.)
   * * * Task n Terminated * * *
d. #n# EXIT PAR F OS [Cr]   (Part filed under current name
                                 and system returns to system
                                 level.)

e. n> LOGOUT [CR]
   * * * Task n Terminated * * *

2.2.5 Part Setup

There is a philosophical difference involved in creating a CADDS 4
data base, in that prior to creating geometry, a *part*, a *drawing*, and
a *view* must be established.

> *Part/model*: the two- or three-dimensional object constructed by
> the designer. Models cannot be seen, except by views in a draw-
> ing.
> *Drawing*: a data base construct necessary to display model geo-
> metry. A part must have at least one drawing, but there is no
> upper limit on the number of drawings that a part may have.
> *View*: a "window" through which part model geometry can be seen.
> Each drawing can contain as many views as needed.
> *Initiate part setup*: A simplified part setup procedure is presented
> here. See information on individual commands for additional
> modifiers, and other details.

In logging in, the system is now at the graphics-level prompt.

#n# ACT PAR (unique part name, 60 characters maximum) (modifiers for new part only) [CR]
Entering New Part (System Response)

#n# ACT DRA (unique drawing name 15 characters maximum) (modifiers for new drawing only) [CR]

#n# ECH FRA [CR] (To display the limits or extents of the drawing.)

#n# DEF VIEW (unique name, maximum 20 characters) : *Draw loc* d_1 *Draw loc* d_2 *Draw loc* d_3 [CR]
d_1 Defines origin of view, X0Y0Z0.
$d_1 d_2$ Diagonals define viewing window.

#n# (User now able to create geometry by inserting entities.)

2.2.6 Part Setup Commands

The following material explains part setup: how to specify and apply the commands required to enter into and create a new part or to recall an old part.
The commands are:

ACTIVATE PART
ACTIVATE DRAWING
SELECT MODE
ECHO FRAME (optional)
DEFINE VIEW

ACTIVATE PART

This command is used to create a new or recall an old part. In creating a new part, initial conditions can be specified with modifiers.

Syntax

#n# ACT PAR unique part name (modifiers) [CR]
Part name restricted to 60 characters with no spaces and cannot end with a period (.)

Modifiers (Used for New Parts Only)

MUNITn   To specify dimension units for the model:
      n =   MM, CM, M, KM, IN (default), FT, MI.
DIMn   Specifies whether the part will be created in two dimensions or in three dimensions:
      n =   DIM 2D, DIM 3D (default).

Example 2

#n# ACT PAR CRANKSHAFT MUNIT FT DIM 2D [CR]

Creating new part named CRANKSHAFT. Measuring units are feet and drawing is in two dimensions.

ACTIVATE DRAWING

This command is used to activate a different drawing, either old or new.

Old: Drawing displayed zoomed to extent of screen.
New: Size parameters established with modifiers or default values.

Syntax

#n# ACT DRA unique drawing name (modifiers) [CR]
    Drawing name restricted to 15 characters.

Modifiers (Used for New Drawings Only)

HGTn   Specify explicit height and width of drawing when plotted
WDTn   true size.
UNITn   Designates units for height, width, and drawing coordin-
         ate system. Units are IN (default), FT, MI, MM, CM,
         and KM.
SIZEn   Specify drawing size: n =

Letter	Height		Width
A	8.5	X	11 in.
B	11	X	17 in.
C	17	X	22 in. (default)
D	22	X	34 in.
E	34	X	44 in.

Example 3

To activate a new drawing:

a. #n# ACT DRA DETAIL.1 UNIT MM [CR]
   Drawing named *Detail.1*; units are millimeters.
b. #n# ACT DRA DETAIL.2 SIZE B [CR]
   Drawing named *Detail.2* on B size (11 X 17 in.) sheet.
c. #n# ACT DRA DETAIL.3 [CR]
   Drawing named *Detail.3*; units are inches on C-size drawing.

Example 4

To activate (recall) existing drawing—no modifiers:

#n# ACT DRA DETAIL.1 [CR]

Change drawing size. The ENLARGE DRAWING command allows the user to enlarge an existing drawing. The part in which the drawing resides must be active, but the drawing to be enlarged cannot be active.

#n# ENLarge DRAW Drawing Name (mod) [CR]

Modifiers

#n# LIST DRA (mod) [CR]   Lists active drawing name.
Modifier—ALL   Lists all drawing names in active part
                           with asterisk noting active drawing
                           name.

ECHO FRAME

This command makes visible or displays the border of the extents of the command drawing sheet within which the geometry will be created (Fig. 2.4). This command is optional and is not required prior to entering graphics commands.

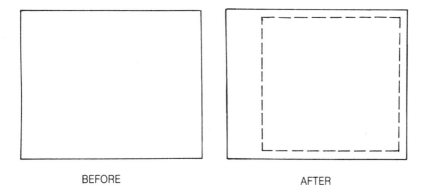

BEFORE   AFTER

FIGURE 2.4

Syntax

#n# ECH FRA [CR]

Select Mode

Mode refers to Model or Draw mode. Model mode is default and is used to define the model by the designer. Draw mode is used by the drafter to create from the model the final specifications, including dimensions and notes.

Syntax

#n# SEL MODE MODEL [CR]
#n# SEL MODE DRAW [CR]

Define View

There are different strategies for creating a two- or a three-dimensional model; this material is for a two-dimensional model.

This command provides for creating a viewing or clipping window within the currently active drawing to create and display model graphics.

Syntax

#n# DEF VIE viewname (modifiers) : Draw loc d_1 [CR]
$\quad$ d_1 $\quad$ Model origin X0Y0.

Modifiers

SCALEn $\qquad$ There are three methods of expressing scale of view:

SCALE 2.5	Single numeric value
SCALE 5 to 2	Ratio of undimensioned numbers
SCALE 10 IN TO 1KM	Ratio of different units: default 1 in. = 1 in. (This is scaling view units to model units.)

Example 5

Explanation

a. #n# DEF VIE VIEW.A SCALE 0.5 : *Draw loc* d_1 [CR] (Fig. 2.5) Creating new view named VIEW.A. View is at a scale of 0.5. d_1 is the origin of X0Y0 for the coordinates. The view/clipping window of the view is coincident with the viewing extents.

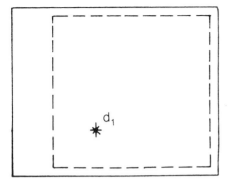

FIGURE 2.5

2.2.7 CRT Status Display

In the upper left-hand corner of the CRT (Fig. 2.6), the status of the following information is displayed for the active part:

 Level: graphics or system
 Part name and drawing name
 Mode: Model or Draw
 Active view or construction plane
 Units chosen: inches, meters, etc.

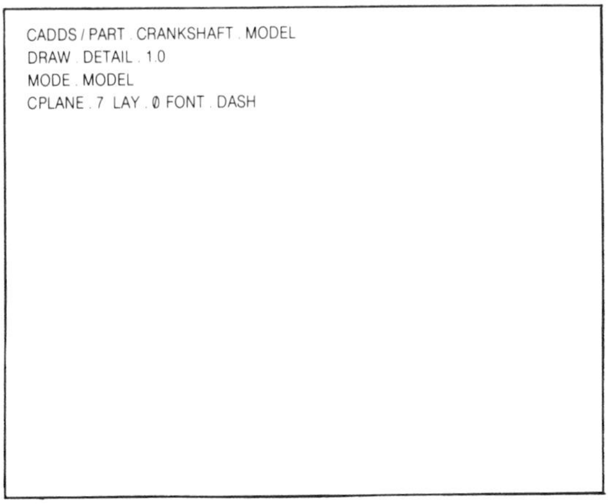

```
CADDS / PART . CRANKSHAFT . MODEL
DRAW . DETAIL . 1.0
MODE . MODEL
CPLANE . 7  LAY . 0  FONT . DASH
```

FIGURE 2.6

2.3 CREATING AND EDITING ENTITIES

2.3.1 Creating Entities

Creating graphics using CADDS 4 with the INSERT commands. The commands are:

INSERT LINE	INSERT ARC	INSERT B-SPLINE
INSERT CIRCLE	INSERT FILLET	etc.

The modifiers specifying diameter and radius are:

DIAMn   To specify diameter.
RADn   To specify radius.

All other modifiers are the same; although the spelling may differ (i.e., PERP instead of PRP for perpendicular) and there are some additional modifiers.
Another change is in data input.

#INS LIN: *Model loc* END d_1 LOC IX_nIY_n [CR].
          Entity mask   Entity mask END, removed by entering
                         loction prior to entering incremental
                         coordinates.

2.3.2 Editing Entities

This section covers the commands that remove or change an existing entity and includes the following commands:

DELETE ENTITY	TRIM ENTITY
EDIT FILLET	DIVIDE ENTITY
CHANGE CIRCLE	

DELETE ENTITY

The DELETE ENTITY command deletes graphical entities from all drawings of a part, both model and drawings. Model entities must be deleted while in Model mode and Draw entities while in Draw mode.

Mode

Both.

Syntax

#n# DEL ENT : *Model ent* d_1 [CR]
      d_1 Model entity deleted.

Editing or changing an existing entity is preferable to deleting it and entering a new one because it avoids gaps in the data storage on the disk.

EDIT FILLET

The EDIT FILLET command is used to change the radius of an existing fillet.

Mode

Both.

Syntax

#n# EDIt FIL RADn : *Model ent* d_1 *Model ent* d_2d_3 [CR]
    RADn   Specifies size of new radius.
    d_1 Identifies fillet to be changed.
    d_2d_3 Identify entities fillet connects in counterclockwise
            direction.

Example 1

#n# EDIt FIL RAD3 : *Model ent* d_1 *Model ent* d_2d_3 [CR] (Fig. 2.7).

CHANGE CIRCLE

The CHANGE CIRCLE command is used to change the radius or diameter of an existing circle.

Mode

Both.

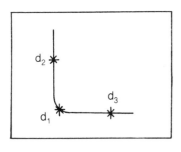

BEFORE

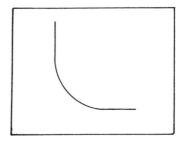

AFTER

FIGURE 2.7

Syntax

#n# CHAnge CIR mod : *Model ent* d_1 [CR]
     d_1 Identifies the circle to be changed.

Modifiers

RADn   "n" equals new radius of circle.
DIAMn   "n" equals new diameter of circle.

Example 2

Explanation

#n# CHA CIR DIAM4 : *Model ent* d_1 [CR] (Fig. 2.8)

The TRIM and DIVIDE ENTITY commands provide for editing existing
entities in the part data base.

TRIM ENTITY

The TRIM ENTITY command is used to shorten or lengthen an existing entity by relocating the endpoints. The noun is ENTITY rather
than the type of entity (line, circle, etc.) as in CADDS 3. Entities
may be lines, circle, arcs, fillets, or B-splines.

Syntax

#n# TRIm ENT mod : *Model ent* d_1 *Model loc* $d_2 d_3$ [CR]
     d_1 End of entity to be relocated.
     d_2 New location of endpoint.
     d_3 OPTIONAL—location of other end of entity Prompts and
          digitizing vary with modifiers.

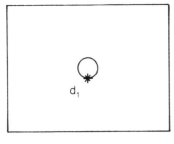

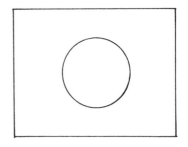

BEFORE   AFTER

FIGURE 2.8

Modifiers

INTOF   Trims specified entity to its intersection with another en-
        tity. Intersection may be real or apparent.

LNGn   Changes length of line an incremented length n. Positive
        value will shorten; negative value will lengthen.

Example 3

Explanation

#n# TRI ENT : *Model ent* d_1 *Model Loc* d_2 [CR] (Fig. 2.9)
    d_1 End of entity to be trimmed.
    d_2 New endpoint location.

Explanation

Line shortened.

b. #n# TRI ENT : *Model ent* d_1 *Model loc* d_2 [CR] (Fig. 2.10)

Explanation

Line lengthened.

c. #n# TRI ENT : *Model ent* d_1 *Model loc* d_2 *Model loc* d_3 [CR] (Fig.
    2.11)

Explanation

Same as preceding example, except that line has been lengthened
at both ends by two digitizes, d_2 and d_3.

d. #n# TRI ENT INTOF : *Model ent* d_1 *Model ent* d_2 [CR] (Fig. 2.12)

Explanation

Line trimmed to intersecton of existing entity.

e. #n# TRI ENT LNG 4 : *Model ent* d_1 [CR] (Fig. 2.13)

Explanation

One digitize required; defines line to be changed in length and end
to be relocated. Length of line changed to 4 units long.

f. #n# TRI ENT 1LNG-2 : *Model ent* d_1 [CR] (Fig. 2.14)

Explanation

Line lengthened by 2 units.

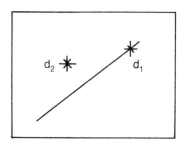
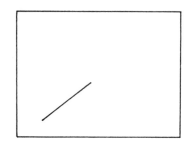

BEFORE   AFTER

FIGURE 2.9

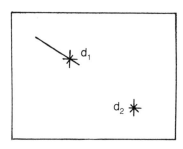
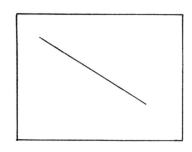

BEFORE   AFTER

FIGURE 2.10

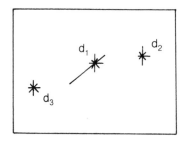
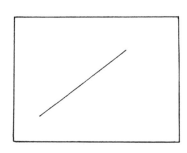

BEFORE   AFTER

FIGURE 2.11

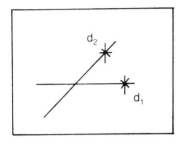

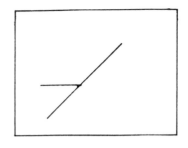

BEFORE

AFTER

FIGURE 2.12

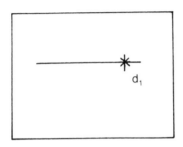

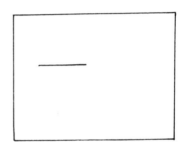

BEFORE

AFTER

FIGURE 2.13

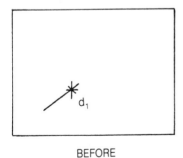

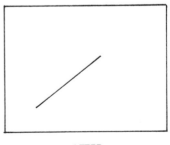

BEFORE

AFTER

FIGURE 2.14

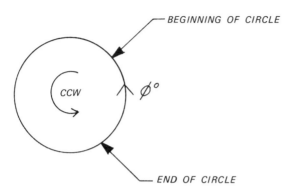

FIGURE 2.15

It must be remembered that a circle is a curved line created in a counterclockwise rotation, beginning at 0° and ending at 360° (Fig. 2.15). In trimming a circle, as in trimming a line, the ends are relocated.

d_1	Identifies circle and end to be relocated.
d_2	Specifies new location of end being relocated.
d_3 (optional)	Specifies new location of other end of circle.

Example 4

a. #n# TRI ENT : *Model ent* d_1 *Model loc* d_2 [CR] (Fig. 2.16)

   Explanation

   Beginning relocated to d_2. Other end not changed.

b. #n# TRI ENT : *Model ent* d_1 *Model loc* d_2 [CR] (Fig. 2.17)

   Explanation

   End relocated to d_2. Beginning not changed.

c. #n# TRI ENT : *Model ent* d_1 *Model loc* d_2 *Model loc* d_3 [CR] (Fig. 2.18)

   Explanation

   Beginning relocated to d_2. End relocated to d_3.

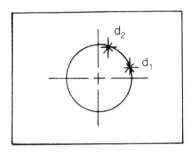

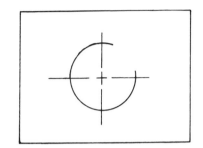

BEFORE

AFTER

FIGURE 2.16

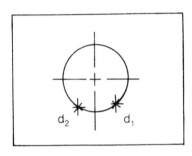

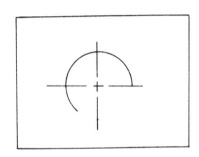

BEFORE

AFTER

FIGURE 2.17

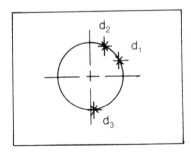

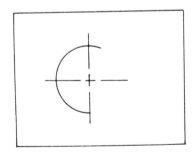

BEFORE

AFTER

FIGURE 2.18

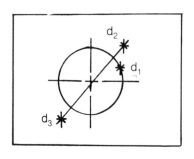 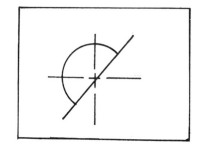

BEFORE   AFTER

FIGURE 2.19

d. #n# TRI ENT INTOF : *Model ent* d_1 *Model ent* d_2 *Model ent* d_3 [CR]
    (Fig. 2.19)

Explanation

Beginning relocated to intersecting entity at d_2. End relocated to intersecting entity at d_3.

DIVIDE ENTITY

The DIVIDE ENTITY command divides a specified entity into separate, smaller entities of the same font. Applies to

    Lines   B-splines
    Arcs   Circles

Mode

Both.

Syntax

#n# DIV ENT mod : *Model ent* d_1 *Model loc* d_2 $\cdots$ d_n [CR]
        d_1 Identifies entity to be divided.
        $d_2 \cdots d_n$ The points at which the entity is divided.

Modifiers

    NDIVn   Number of equal divisions into which entity will be divided;
             one digitize required.
    INTOF   Specified entity is divided at points defined with digitizes
             at intersecting entities.

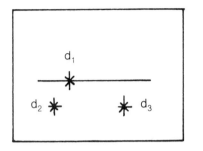 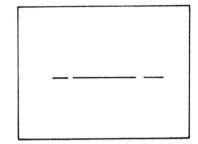

BEFORE

AFTER

FIGURE 2.20

Example 5

a. #n# DIV ENT : *Model ent* d₁ *Model loc* d₂d₃ [CR] (Fig. 2.20)

   Explanation

   One line divided into three shorter lines. Breaks shown for clarification only.

b. #n# DIV ENT NDIV3 : *Model ent* d₁ [CR] (Fig. 2.21)

   Explanation

   One line divided into three shorter lines of equal length.

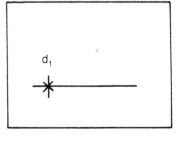 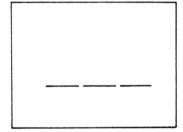

BEFORE

AFTER

FIGURE 2.21

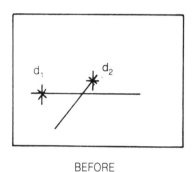

BEFORE   AFTER

FIGURE 2.22

c. #n# DIV ENT INTOF : *Model ent* d_1 *Model ent* d_2 [CR] (Fig. 2.22)

   Explanation

   One line divided into two shorter lines at intersecting entity.

d. #n# DIV ENT INTOF : *Model ent* d_1 *Model ent* d_2 *Model ent* d_3
    Model ent d_4 [CR] (Fig. 2.23)

    #n# DEL ENT : *Model ent* d_5 [CR]

   Explanation

   One line divided into three shorter lines at intersecting entities
and one segment deleted.

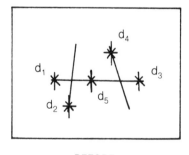

BEFORE

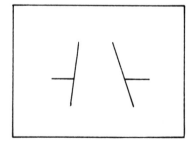

AFTER

FIGURE 2.23

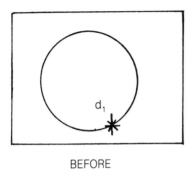 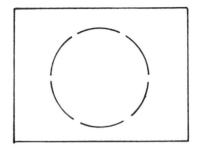

BEFORE   AFTER

FIGURE 2.24

e. #n# DIV ENT NDIV6 : *Model ent* d1 [CR] (Fig. 2.24)

Explanation

One circle divided into six equal arcs.

f. #n# DIV ENT INTOF : *Model ent* d_1 *Model ent* d_2 [CR] (Fig. 2.25)
    #n# DEL ENT : *Model ent* d_3 [CR]

Explanation

One circle divided into two arcs at the two intersections of the line. One arc deleted.

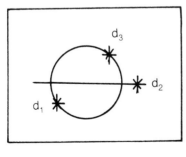 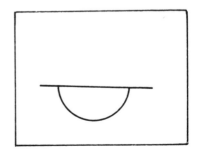

BEFORE   AFTER

FIGURE 2.25

2.4 LINE FONTS, CENTERLINES, CROSSHATCHING, AND GRIDS

2.4.1 Line Fonts

The default for line fonts is solid, but the system does provide for having other line fonts user defined with the SELECT/CHANGE APPEARANCE commands. Line fonts applicable to mechanical design are:

Dash   ----------------------
Phantom   —— -- —— -- ——
Arrow   ————————————➤
Solid   ————————————➤   Default

SELECT APPEARANCE/INSERT LINE FONT

The SELECT APPEARANCE command with the proper modifiers establishes a new line font parameter. The line font name appears as part of the part status in the upper left corner of the screen.

Syntax

#n# SEL APP FONT mod [CR]

Modifiers

DASH
PHANTOM
ARROW

The INSERT LINE FONT command is required to insert a line of the selected font.

Syntax

#n# INS LIN FONT : *Model Loc* d_1d_2 [CR]

Example 1

a. #n# SEL APP FONT DASH [Cr]
Establishes dash line font. Font name appears in part status on screen.
b. #n# INS LIN : *Model loc* d_1d_2 [CR]
New line is solid.
c. #n# INS LIN FONT : *Model loc* d_3d_4 [CR]
New line is dashed.
d. #n# INS CIR DIAM 2 FONT : *Model loc* d_5 [CR] (Fig. 2.26)
Circle is dash font.
e. #n# SEL APP FONT PHANTOM [CR] (Fig. 2.26)

Explanation

New line will be phantom font.

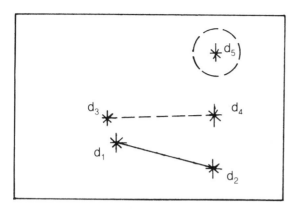

FIGURE 2.26

CHANGE APPEARANCE

The CHANGE APPEARANCE command allows the operator to change the font of an existing entity.

Syntax

#n# CHA APP FONT mod : *Model ent* d_1 [CR]

Modifier

Name of new font: DASH, ARROW, PHANTOM, SOLID. d_1 indicates entity to have new font.

2.4.2 Centerlines

The INSERT CENTERLINE command provides for inserting centerlines, which are displayed as alternate long and short line segments. By default, centerlines are Draw entities, but with the modifier MODEL can be entered as Model entities.

Syntax

#n# INS CLIN mod : *Draw org* d_1 or *Draw ent* d_1 [CR]
      Mode prompt varies depending on the modifier.

Modifiers

POI   Point to point (default).
CIR   Provides two perpendicular centerlines through the center of a circle.
DIAMn   Constructs circular centerline defined by center and diameter.

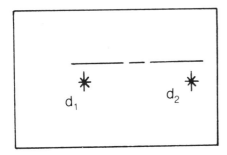

FIGURE 2.27

RADn   Constructs circular centerline defined by center and radius.

CCIR   Constructs a circular centerline defined by three points.

ANGn   Constructs centerline through a point at a specified angle.

MODEL   Inserted as Model entity (default is Draw).

Example 2

a.   #n# INS CLIN : *Draw org* $d_1 d_2$ (Fig. 2.27)

Explanation

Point to point centerline created.

b.   #n# INS CLIN CIR : *Draw ent* d_1 [CR] (Fig. 2.28)

Explanation

Two perpendicular centerlines created through center of existing circle.

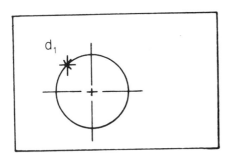

FIGURE 2.28

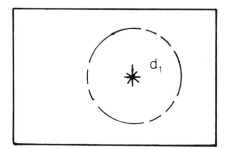

FIGURE 2.29

c. #n# INS CLIN DIAM 2 : *Draw org* LOC d_1 [CR] (Fig. 2.29)

Explanation

Circular centerline with diameter of 2 in. created.

2.4.3 Crosshatching

The CROSSHATCHING command provides for defining a cross-sectioned surface (Fig. 2.30) by inserting nonindependent lines within the boundaries of that surface. Crosshatching is a Draw entity by default, but with modifier MODEL can be entered as a Model entitity. Semicolons are entered in data input to define separate boundaries.

Syntax

#INS XHA mod : *Draw/Model ent* $d_1 \cdots d_n$; *Draw/Model ent* $d_x \cdots d_y$
     [CR]

$d_1 \cdots d_n$	Boundary 1.
;	Semicolon as a separator between boundaries.
$d_x \cdots d_y$	Boundary 2.

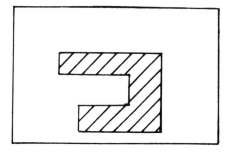

FIGURE 2.30 Cross-sectioned surface.

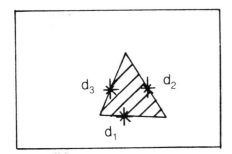

FIGURE 2.31

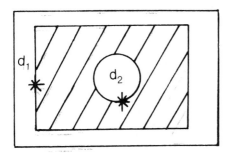

FIGURE 2.32

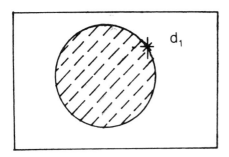

FIGURE 2.33

Modifiers

ANGLEn	Angle of crosshatching (default is 45°).
SPACINGn	Spacing or distance between lines (default is 0.236 in.).
FONTn	Lines in any appearance font (default is solid). Fonts are available to represent specific material.
MODEL	Inserted as model entity (default is Draw entity).

Restrictions: Maximum number of boundaries is 100; maximum number of entities per boundary is 350.

Example 3

a. #n# INS XHA : *Draw/Model ent* $d_1d_2d_3$ [CR] (Fig. 2.31)

Explanation

Solid lines added at 45° and 0.236 in. apart.

b. #n# INS XHA ANGle 60 SPACING 0.5 : *Draw/Model ent* CHN d_1; *Draw/Model ent* d_2 [CR] (Fig. 2.32)

Explanation

Solid lines added at 60° and 0.5 in. apart. Note semicolon separator between boundaries.

c. #n# INS XHA FONT DASH : *Draw/Model ent* d_1 [CR] (Fig. 2.33)

Explanation

Dash lines add at 45° and 0.236 in. apart.

2.4.4 Grids

The grid commands, SELECT GRID and ECHO GRID, provide visible grid points on the screen which can be used for reference only, or the grid may be active, which means that only grid points can be digitized. Grids are applicable to both Model and Draw modes.

SELECT GRID

The SELECT GRID command provides for specifying the mode and grid point spacing.

Syntax

#n# SEL GRI mod [CR]   or   : *Model loc* d_1 [CR]

Modifiers

MODEL	Model mode grid—default.
DRAW	Draw mode grid.
REC	Rectangular spacing—default.
DXn	Horizontal point spacing; n. Default is 1 in.
DYn	Vertical point spacing; n. Default is 1 in.
DGn	Horizontal and vertical point spacing; n. Default is 1 in.
ANGn	Grid angle, n. Default is 0 in.
ORIGIN	Specify origin of grid by digitizing.

ECHO GRID

The ECHO GRID command provides for making the selected grid visible and either active or inactive.

Syntax

#n# ECH GRI mod [CR]

Modifiers

ON	Grid becomes visible. Default.
OFF	Grid becomes invisible. Invisible grid cannot be active.
SNAP	Grid becomes active, digitizes snaps to grid points. Default.
NOS	No snap, grid points inactive. Reference only.

Example 4

a. #n# SEL GRI DG.5 [CR]
   Specify grid with rectangular spacing and horizontal and vertical spacing of 0.5 in.
b. #n# ECH GRI ON NOS [CR]
   Grid becomes visible but inactive. Reference only.

LIST GRID

The LIST GRID command provides the status of the existing grid.

Syntax

#n#LIST GRID [CR]
     **Default Parameter **
         TYPE   -----
         DIMENSION   -----
         DELTA ANG   -----
         DELTA SP   -----
         ORIGIN   -----
         ORIENTATION MATRIX   -----

2.4.5 Text

Insert Text

Inserting text in CADDS 4 should present a minimum of problems. Additional modifiers are provided and a partial list of them is included in this material.

Mode. Text can be entered as either a Model or a Draw entity depending on the mode of the system upon insertion. An example of text as a Draw entity is when it is a note on the drawing, such as FINISH ALL OVER. An example of text as a Model entity is when it is a feature of the model, such as lettering on a nameplate.

Syntax

#n# INS TEX / text goes here/mod : *Draw loc* d_1 [CR]
   Mode prompt varies with modifiers and mode of system.

Modifiers (Partial List)

In addition to the parameter modifiers (font, height, etc.), CADDS 4 provides modifiers for positioning text relative to an existing entity.

PRL   Text entered parallel to existing specified line.
PRP   Text entered perpendicular to existing specified line.
CURV   Text entered relative to existing specified arc or circle.
FITL   Text size automatically changed to fit within two digitized
          locations.

Example 5

a. #n# SEL MODE MODEL [CR]
    #n# INS TEX /PART NO. CR
       123456 / PRL HGT.25 : *Model ent* d_1 *Model loc* d_2 [CR]
          (Fig. 2.34)

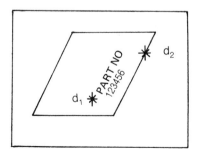

FIGURE 2.34

Explanation

Text is entered as a Model entity parallel to the Model entity line digitized d_1, at a location d_2, and at a height of 0.25 in.

b. #n# SEL MODE DRAW [CR]
    #n# INS TEX /SEE NOTE 3/ CURV : *Draw ent d_1 Draw loc d_2* [CR]
    (Fig. 2.35)

Explanation

Arc identified with d_1 must be a Draw entity.

CHANGE TEXT

The CHANGE TEXT command isused to change only the parameters of existing text.

Mode

Both.

Syntax

#n# CHA TEX mod : *Model/Draw ent d_1* [CR]

Parameter Modifiers (Partial list)

HGTn   Change height to n value.
FONTn   Change font to new font number specified.
LJT   Left justification.
RJT   Right justification.
CJT   Center justification.

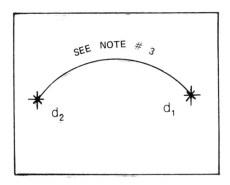

FIGURE 2.35

2.4.6 Dimension

Insert Dimension

Dimensioning in CADDS 4 provides many modifiers. A partial list of them is provided in Section 2.9.

Mode

Dimensions can be entered in either mode, but they are always Draw entities. The system must be in the Draw mode to delete a dimension.

Syntax

#n# INS LDI mod : *Draw/Model ent-end*   or   *Model/Draw ent-ent*
       d_1d_2 *Draw loc* d_3 [CR]

#n# INS DDIM mod : *Draw/Model ent* or *Model/Draw-ent* d_1 *Draw*
       loc d_2 [CR]

Model prompt varies depending on type of dimension and active mode of system.

Modifiers (Partial List)

Modifiers are grouped in a hierarchical structure of associated modifiers.

Modifier		Level
TEXT	Controls text characteristics.	1
HEIn	Dimension height. Default is 0.156 in.	2
MAIN "t"	Dimension is replaced by new text "t".	2
APP "t"	Given text "t" is appended to the main text.	2
PRE "t"	Given text "t" prefixes main text.	2
TOL	Controls tolerance characteristics.	1
POSn	Positive tolerance value is n.	2
NEGn	Negative tolerance value is n.	2
BOTHn	Both positive and negative values are n.	2
PRECn	Specifies number of decimal places for tolerance values. Default is 3 in.	2

Example 6

a. #n# INS ldim tex hei .25 TOL BOTH.003 : *Model/Draw ent-end* d_1d_2
   Draw loc d_3 [CR] (Fig. 2.36)

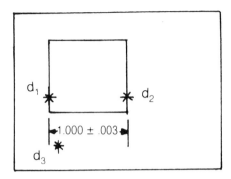

FIGURE 2.36

Explanation

Dimension text size is 0.25 in. high and tolerance of 0.003 in. is added as both positive and negative values.

b. #n# INS DDIM TEX APP /DIA./ : *Model/Draw ent* d_1 *Draw loc* d_2 [CR] (Fig. 2.37)

Explanation

Text DIA. is appended to the basic dimension.

Changing Dimensions and Tolerances

The CHANGE DIMENSION command provides for revising the attributes of an existing dimension. The sequence of entering modifiers is

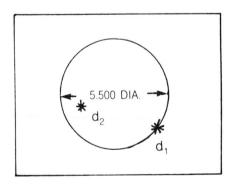

FIGURE 2.37

hierarchical and the same modifiers used in inserting dimensions and tolerances apply to changing them.

Mode

Both.

Syntax

#n# CHA DIM mod : *Model/Draw ent* $d_1 \cdots d_n$ [CR]
   $d_1 \cdots d_n$ Identify dimensions affected.

#n# CHA DIM mod : *Model/Draw ent* d_1 *Draw loc* d_2d_3 [CR]
   d_1 Identifies dimension affected.
   d_2d_3 Used with ORIgin modifier to define vector.

Modifiers (Partial List)

Modifier		Level
TEX	Text characteristic to be affected.	1
HEIn	Dimension is changed to height n.	2
ORI	Dimension location is changed.	2

Example 7

a. #n# CHA DIM TEX HEI.24 : *Model/Draw ent* d_1 [CR] (Fig. 2.38)

Explanation

Height of existing dimension text digitized (d_1) is changed to 0.25 in.

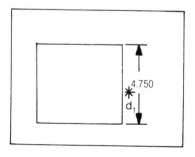
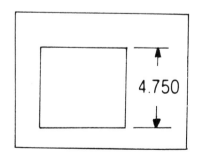

BEFORE   AFTER

FIGURE 2.38

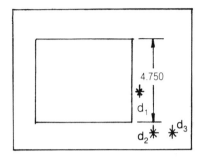

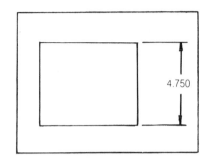

BEFORE   AFTER

FIGURE 2.39

b. #n# CHA DIM TEX ORI : *Model/Draw ent* d_1 *Draw loc* d_2d_3 [CR]
    (Fig. 2.39)

Explanation

Origin (location) of existing dimension digitized (d_1) is relocated
by a vector defined by d_2 and d_3.

2.5 DISPLAY CONTROL: ZOOM AND SCROLL DRAWING

2.5.1 Zoom Drawing

The ZOOM DRAWING command permits the user to increase or reduce
the graphics displayed on the screen. Both model and drawing entities are affected.

Syntax

#n# ZOOm DRAw (mod)   or   : *Draw loc* d_1d_2 [CR]

Modifiers

IN   The drawing is increased by twice its present size.
OUT   The drawing is reduced by one-half its present size.
Rn   Used with modifiers *in* or *out*, the drawing is redisplayed
       by a factor of n from its present size.
An   The drawing is redisplayed by a factor of n of its true size.
ALL   The drawing is both zoomed and scrolled so as to fit on the
       screen.
WIN   A window defined by two diagonal digitizes is zoomed and
       scrolled to fill the screen.

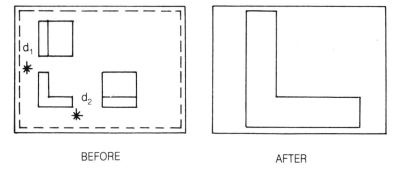

BEFORE

AFTER

FIGURE 2.40

Example 1

a. #n# ZOO DRA WIN : *Draw loc* d_1d_2 [CR] (Fig. 2.40)
b. #n# ZOO DRA ALL [CR] (Fig. 2.41)

2.5.2 Scroll Drawing

The SCROLL DRAWING command permits the user to shift the displayed image on the screen. Scrolling displacement is indicated with two digitizers; the first digitized location is shifted to the second digitized location.

Syntax

#n# SCRoll DRAw : *Draw loc* d_1d_2 [CR]

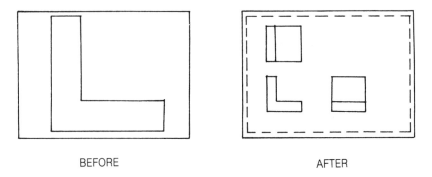

BEFORE

AFTER

FIGURE 2.41

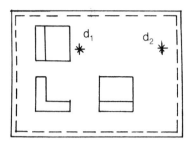

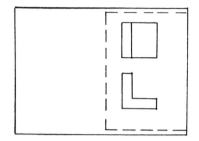

BEFORE   AFTER

FIGURE 2.42

Example 2

a. #n# SCR DRA : *Draw loc* d_1d_2 [CR] (Fig. 2.42)
b. #n# ZOOm DRAw WIN : *Draw loc dd* [CR]
    #n# SCRoll DRAw : *Draw loc* d_1d_2 (Fig. 2.43)

2.5.3 Scroll View

SCROLL VIEW repositions the drawing on the screen. SCROLL VIEW
repositions the view within the drawing.

Syntax

#n# SCRoll VIEw : *View* $d_1 \cdots d_n$; *Draw loc* d_xd_y [CR]
    $d_1 \cdots d_n$ Identifies view(s) affected.
    ;   Enter semicolon to switch to location mode.
    d_xd_y Scrolling displacement from d_x to d_y within the
                 clipping of the view.

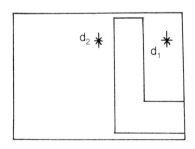

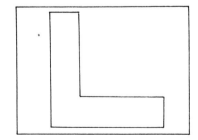

BEFORE   AFTER

FIGURE 2.43

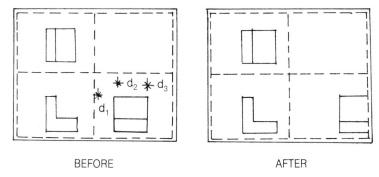

BEFORE   AFTER

FIGURE 2.44

Example 3

#n# SCRoll VIEw : *View* d_1; *Draw loc* d_2d_3 [CR] (Fig. 2.44)

2.6 GRAPHICS MANIPULATION: MIRROR, TRANS-LATE, AND ROTATE ENTITIES

2.6.1 Mirror Entity

A mirror image of the existing geometry is generated about a defined axis.

Syntax

#n# MIR ENT mod : *Model ent* $d_1 \cdots dn$; *Model loc* d_xd_y [CR]

$d_1 \cdots d_n$	Specify entities to be mirrored.
;	Semicolon as separator to change to locational input.
d_xd_y	Define mirroring axis with two digitizes.

Modifiers

COPY	Original geometry not affected and mirror image provided. Default is not to copy. The copy will be made on the current active layer.
SAMLAY	The mirrored geometry remains on the same layer as the original geometry. Default.
COPY SAMLAY	The mirrored copy will be on the same layer as the original geometry.

Example 1

#n# MIR ENT COPY : *Model ent* WIN d_1d_2; *Model loc* END d_3d_4 [CR] (Fig. 2.45)

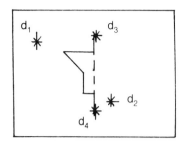
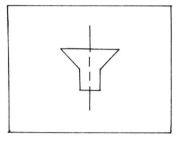

BEFORE   AFTER

FIGURE 2.45

2.6.2 Translate Entity

Existing geometry is moved in a linear manner as defined by a vector.

Syntax

#n# TRAN ENT mod : *Model ent* $d_1 \cdots d_n$; *Model loc* $d_x d_y$ [CR]

$d_1 \cdots d_n$ Specify entities to be moved.

;   Semicolon as separator to change to locational input.

$d_x d_y$ Two digitizes required to define vector, direction, and magnitude of move.

More than one copy may be specified by entering second digitizer (d_y) as explicit incremental coordinates from first digitize (d_x) and total number of copies required. d_x IX 3N5.

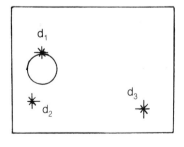
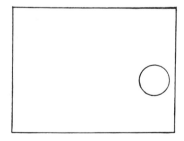

BEFORE   AFTER

FIGURE 2.46

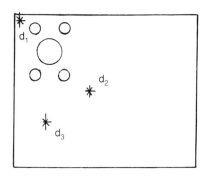 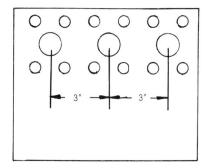

BEFORE   AFTER

FIGURE 2.47

Modifiers

COPY   A copy of the original geometry is moved to the
                   new location on the current active layer. De-
                   fault is no copy.

SAMLAY   The moved geometry is on the same layer as the
                   original geometry.

COPY SAMLAY   The moved copy is on the same layer as the orig-
                   inal geometry.

Example 2

a. #n# TRAN ENT : *Model ent* d_1; *Model loc* d_2d_3 [CR] (Fig. 2.46)
b. #n# TRAN ENT COPY : *Model ent* WIN d_1d_2; *Model loc* d_3 1X3 N2
    [CR] (Fig. 2.47)

2.6.3 Rotate Entity

Existing geometry is rotated about a point.

Syntax

#ROT ENT mod : *Model ent* $d_1 \cdots d_n$; *Model loc* d_x [CR]
    $d_1 \cdots d_n$ Specify entities to be rotated.
    ;   Semicolon is separator to change to locational input.
    d_x One digitizer defines center point of rotation.

Modifiers

COPY   A copy of the original geometry is rotated to the
                   new location on the current active layer. De-
                   fault is no copy.

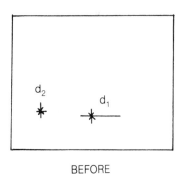
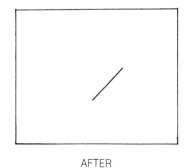

BEFORE   AFTER

FIGURE 2.48

SAMLAY   The rotated geometry is on the same layer as the
                 original geometry.
COPY SAMLAY   The rotated copy is on the same layer as the orig-
                 inal geometry.
ANGn   Specific angle of rotation in degrees about Z axis.

Example 3

a. #n# ROT ENT ANG 45 : *Model ent* d_1; *Model loc* d_2 [CR] (Fig.
    2.48)
b. #n# ROT ENT ANG 45 COPY : *Model ent* d_1; *Model loc* d_2 POI d_3
    [CR] (Fig. 2.49)

Explanation

Additional copies may be generated by specifying the incremental
axis and number of additional copies required IZ0 Nn.

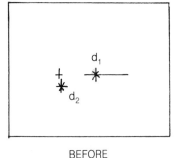
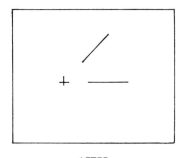

BEFORE   AFTER

FIGURE 2.49

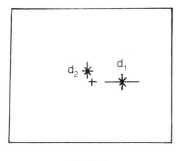

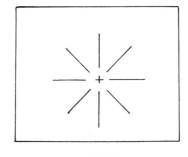

BEFORE   AFTER

FIGURE 2.50

c. #n# ROT ENT ANG 45 COPY : *Model ent* d$_1$; *Model loc* PO1 d$_2$ *loc*
    IZ0 N6 [CR] (Fig. 2.50)

2.7 LAYERS

The system has 255 layers numbered 0 through 254.

SELECT LAYER

The SELECT LAYER command determines the active construction
layer. Entities subsequently inserted will be on the selected layer
number. Only one layer can be active at a time.

Syntax

#n# SEL LAY n [CR]
    Construction layer number appears in part status on the screen.

ECHO LAYER

The construction layer is always visible, and with the ECHO LAYER
command other layers, as specified, can also be made visible.

Syntax

#n# ECH LAY mod [CR]

Modifiers

n	Operator-specified layers made visible; numbers can be in string form: 0 4 8 10 or in from-to form: 0-6 8-12; spaces between a string of numbers, hyphen between from-to.
ALL	All 255 layers made visible.
DRAW	Only drawing entities will be visible on specified layers.

CHANGE LAYER

With the CHANGE LAYER command selected entitites are moved from their original layer to the specified layer.

Syntax

#n# CHA LAY n : *Model ent* $d_1 \cdots d_x$ [CR]
    n   Specifies the layer to which the selected entities
                      will be moved.
    $d_1 \cdots d_x$ Identifies entities to be moved.
    Restriction   Entities to be moved must be visible.

COPY ENTITY

The COPY ENTITY command allows for creating a copy of an existing entity at the same coordinates on a specified layer.

Syntax

#n# COP ENT FROM n To m : *Model ent* $d_1 \cdots d_x$ [CR]
    n   Specifies the current layer of the entity to be
                      copies.
    m   Specifies the layer to which the copied entity will
                      be.
    $d \cdots d_x$ Identifies entities to be copied on new layer.

COUNT ENTITIES

The COUNT ENTITY command provides data and status on current active part.

Syntax

#n# COU ENT [CR]

Example response:

Count Entity	LAYERS USED : 0, 10
4 line	GEOM ONLY : 0
4 arc	TEXT ONLY : 10
1 text	

Total = 9

LIST LAYER

This command lists the active construction and/or the currently visible model or drawing layers for the active drawing or views.

Mode

Both.

Syntax

#n# LIST LAY mod [CR]

Modifiers

ACT   Lists the active construction layer.
DRAW   Lists data for the active drawing.
MODEL   Lists data for all views of the model.
ALL   Lists all data. Default.

Example 1

#n# LIS LAY ALL [CR]
    Active construction layer: 100
    Visible layers:
    Drawing: 0-100
    View: 0-100

2.8 DYNAMIC MOVEMENT

2.8.1 Stretch Line

The STRETCH LINE command provides for relocating one end of an existing line while the other end remains fixed. The line appears to be stretched.

Mode

Both.

Syntax

#n# STR LIN : *Model ent* d_1 *MODEL loc* d_2 [CR]
    d_1 Identifies the end of the line to be relocated.
    d_2 Specifies new locations of end of line through digitizing or explicit coordinates.

Example 1

a. #n# STR LIN : *MODEL ent* d_1 *MODEL loc* d_2 [CR] (Fig. 2.51)
b. #n# STR LIN : *MODEL ent* d_1 *MODEL loc* end d_2 [CR] (Fig. 2.52)

2.8.2 Stretch Entity

The STRETCH ENTITY command provides for stretching the ends of lines and translating other entities within a user-defined window.

Mode

Both.

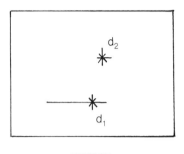

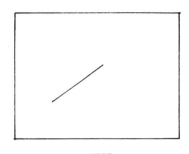

BEFORE   AFTER

FIGURE 2.51

Syntax

#n# STR ENT (mod) : *DRAW loc* d_1d_2; *DRAW loc* d_3d_4 [CR]

d_1d_2 Defines rectangular window and entities affected.

;   Semicolon as delimiter to switch to location mode.

d_3d_4 Defines vector specifying direction and magnitude of displacement.

Modifiers

PWIN   Specifies polygon window will be used. Default is rectangular window.

-TRAN   This modifier allows the lines to be stretched but prevents the translating of entities fully within the window. Default is translate.

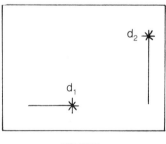

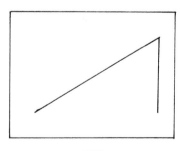

BEFORE   AFTER

FIGURE 2.52

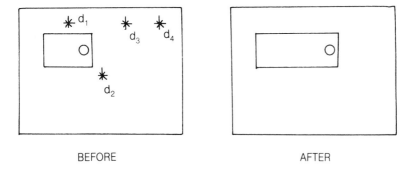

BEFORE   AFTER

FIGURE 2.53

Example 2

#n# STR ENT : *DRAW loc* d_1d_2; *DRAW loc* d_3d_4 [CR] (Fig. 2.53)

2.8.3 Select Dynamics

Explanation

The SELECT DYNAMICS command provides for dynamically stretching and translating entities.

Mode

Both.

Syntax

#n# SEL DYN (mod) [CR]

Modifiers

ON   Turn dynamic digitizing on.
OFF   Turn dynamic digitizing off. Default.

2.8.4 Dynamic Movement

Dynamic movement is initiated by pressing the middle button on the digitizing pen and moving it across the tablet. When the moving image is in the desired position, the bottom button is pressed to set the image in place. After entering either the STRETCH or TRANSLATE command and identifying the affected entities, the mode prompt is

MODEL dyn   or   *DRAW dyn*

Example 3

a. #n# SEL DYN ON [CR]
b. #n# STR LIN : *MODEL ent* d_1 *MODEL dyn* d_2 [CR]
c. #n# TRAN ENT : *MODEL ent* d_1; *MODEL dyn* $d_2 d_3$ [CR]
d. #n# SEL DYN OFF [CR]
e. #STR LIN : *MODEL ent* d_1 *MODEL loc* d_2 [CR]

Note dynamic prompt (*dyn*) when dynamics is turned on.

2.9 DIMENSIONING AND TOLERANCING

2.9.1 Select Dimension

With the SELECT DIMENSION command, parameters are defined that
control subsequent dimension creations. The default values are
changed. The newly established parameters can be overridden using
corresponding modifiers during the individual dimension insertion
command.

The newly established parameters are in effect if entered while in
an active part, for that part only. If entered while not in an active
part, they will be in effect until log-out. Entering CADDS resets all
parameters to default values.

The SELECT DIMENSION command provides the user with many options, from the size and type of arrowheads to the type of drafting
standard used: International Organization of Standardization, Japan-
ese Industrial Standards, or American National Standards Institute
(default). This unit covers selected modifiers.

Mode

Dimensions and tolerances are Draw mode entities but can be entered
in either mode. The mode prompt is *DRAW/MODEL ent* or *DRAW/MODEL
ent-end*. However, the system must be in Draw mode to delete a dimension entity.

Dimensioning

Syntax

#n# SEL DIM mod [CR]

Modifiers (Partial List)

The modifiers are grouped in a herarchical structure of associated
modifiers.

Modifier		Level
UNITS	Specifies the dimensioning units.	1
LINEAR	Controls linear units.	2
PRIMARY	Controls units for primary dimensions.	3
SECONDARY	Controls units for secondary dimensions.	3
BOTH	Controls units for primary and secondary dimensions.	3
INCHES	Dimensions in inches. Default.	4
FEET	Dimensions in feet.	4
MILES	Dimensions in miles.	4
MILLIMETERS	Dimensions in millimeters.	4
CENTIMETERS	Dimensions in centimeters.	4
METERS	Dimensions in meters.	4
KILOMETERS	Dimensions in kilometers.	4
ANGULAR	Controls angular units.	2
DEGREES	Angular dimensions will be in degrees and fraction of degrees. Default.	3
MINUTES	Angular dimensions in degrees, minutes, and fractions of minutes.	3
SECONDS	Angular dimensions in degrees, minutes, seconds, and fractions of seconds.	3
PRECISION	Controls number of decimal places or denominator for fraction format.	1
PRIMARY n	Primary precision is n.	2
SECONDARY n	Secondary precision is n.	2
BOTH n	Primary and secondary precisions are n.	2
ANGULAR n	Angular precision is n.	2
TEXT	Controls text characteristics.	1
FORMAT	Format for dimension.	2
DECIMAL	Dimensions in decimals. Default.	3
FEET & INCHES	Format in feet and inches 1'-10". Use fraction modifier to indicate fraction format.	3
FRACTION	Fraction used. Denominator specified by the PRECISION modifier.	3
LARGE	Numerator and denominator normal height separated by a horizontal bar.	4
LAYER	Specifies layer for dimensions. Default: current layer.	1
PRIMARY n	Primary dimension on layer n. Default.	2
SECONDARY n	Secondary dimensions on layer n.	2
DUAL n	Dual dimensions on layer n.	2
PRIMARY	Controls primary dimensioning status.	1
ON	Turns on. Default.	2
OFF	Turns off.	2

Modifier		Level
SECONDARY	Controls secondary dimensioning status.	1
ON	Turns on.	2
OFF	Turns off. Default.	2
DUAL	Controls dual dimensioning status.	1
ON	Turns on.	2
OFF	Turns off. Default.	2

Note: To change the default value for the height of dimensions (0.156) enter

#n# SEL TEX HEI n [CR]

Example 1

Specifying new dimension parameters may be with a single command or a series of commands.

```
              (1) (2) (3) (4) (3)  (4)           (Modifier level.)
a.  #n# SEL DIM UNI LIN PRI INC SEC MILL         (Specifying pri-
                                                  mary and
                                                  secondary
                                                  units.)

              (1)   (2)   (2)
              PREC PRI 2 SEC 1                    (Specifying de-
                                                  nominator.)

              (1)   (2)    (2)     (2)
              LAY PRI 10 SEC 20 DUAL 30 [CR]      (Specifying
                                                  layer dimen-
                                                  sions on.)

              (1) (2) (1)    (2)
        #n# SEL DIM SEC ON  DUAL ON  [CR]         (Turns secon-
                                                  dary and dual
                                                  dimensions
                                                  on.)
```

#n# INS LDI : *DRAW/MODEL ent-end* $d_1 d_2$ *DRAW loc* d_3 [CR] (Fig. 2.54)

b. #n# SEL DIM UNIts LINear PRImary INCh SECondary MILLimeter PREcision
PRImary 2 SECondary 1 LAYer PRImary 10 SECondary 20 DUAL 30 [CR]

#n# SEL DIM SEC ON DUAL ON [CR]

#n# INS LDIM : *DRAW/MODEL ent-end* $d_1 d_2$ *DRAW loc* d_3 $\cdots$ [CR] (Fig. 2.55)

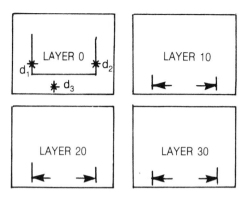

FIGURE 2.54

Tolerancing

In establishing parameters for tolerancing, the basic SELECT DI-MENSION command is used along with the modifiers unique to tolerancing. There are no default parameters for tolerancing. *Default is no tolerancing.*

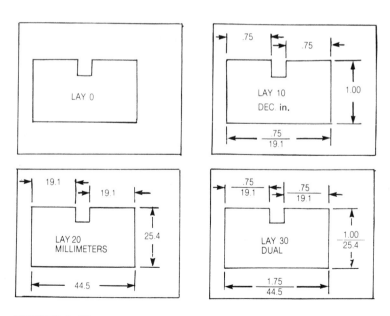

FIGURE 2.55

Syntax

#n# SEL DIM mod [CR]

Modifiers

The tolerancing modifiers, like the dimensioning modifiers, are grouped in a hierarchical structure of associated modifiers.

Modifiers		Level
TOLERANCE	Controls tolerance specification.	1
PRIMARY	Controls primary tolerance.	2
POSITIVE n	Positive tolerance is n. (+0.002)[a]	3
NEGATIVE n	Negative tolerance is n. (-0.003)[s]	3
BOTH	Both values are n. (+0.002)[a]	3
SECONDARY	Controls secondary tolerance.	2
POSITIVE n	Positive tolerance is n. (+0.002)[a]	3
NEGATIVE n	Negative tolerance is n. (-0.003)[a]	3
BOTH n	Both values are n. (+0.002)[a]	3
POSITIVE n	Primary positive tolerance is n. (+0.002)[a]	2
NEGATIVE n	Primary negative tolerance is n. (-0.003)[a]	2
BOTH n	Primary positive and negative tolerances are n. (+0.002)[a]	2
PRECISION	Controls number of decimal places in tolerance.	2
PRIMARY n	Primary tolerance has n decimal places.	3
SECONDARY n	Secondary tolerance has n decimal places.	3
BOTH n	Both tolerances have n decimal places.	3
ANGULAR n	Angular tolerance has n decimal places.	3
TYPE	Controls tolerancing method or form.	2
INCREMENTAL	Tolerance in incremental form, X.XX + 0.XX.	3
DASHED	Tolerance in dashed form, X.XX - X.XX.	3
LIMIT	Tolerance in limit form, X.XX.	3
ANGULAR	Controls the number of decimal places for angular dimension tolerance values.	2
POSITIVE n	Positive tolerance is n.	3
NEGATIVE n	Negative tolerance is n.	3
BOTH n	Positive and negative tolerances are both n.	3
ON	Turns tolerancing on.	2
OFF	Turns tolerancing off, Default.	2

[a]( ) shown as example only.

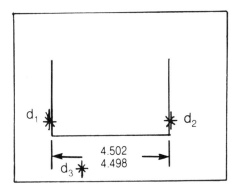

FIGURE 2.56

Example 2

(1) (2)   (2) (3)   (2) (3) (2)

a. #n# SEL DIM TOL BOTH 0.002 PREC BOTH 3 TYPE LIMit ON [CR]

    #n# INS LDI : *DRAW/MODEL ent-end* $d_1 d_2$ [CR] (Fig. 2.56)

b. #n# SEL DIM TOL TYPE INCR [CR]

    #n# INS LDI : *DRAW/MODEL ent* $d_1 d_2$ *DRAW loc* d_3 [CR] (Fig. 2.57)

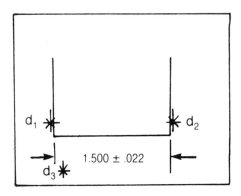

FIGURE 2.57

2.9.2 Tolerance Stack-Up

Purpose

To learn the TOLERANCE STACK-UP command and how to apply it to automatically compute the tolerance stack-up (tolerance accumulation) of selected linear dimensions in the same direction, vertical or horizontal. On mechanical engineering drawings, a tolerance stack-up is performed to determine linear tolerances or clearance through the accumulation of tolerances on mating parts to optimize cost-effectiveness.

General

Three types of stack-ups can be calculated:

1. Arithmetic (add/subtract)
2. Statistical (one standard deviation)
3. Realistic or bender (one and one-half standard deviations)

Note: Dimensions must be in the limit form, maximum/minimum: X.XXX
                                                                X.XXX

Dimensions must be specified as either positive or negative, that is, either added or subtracted from the stack-up computation.

Syntax

#n# CALC TSTack mod:

Modifiers

ARITH   Provide arithmetic stack-up. Default when no modifier specified.
STAT   Provide statistical stack-up. Default when no modifier specified.
REAL   Provide realistic stack-up. Must be specified.

When a modifier is specified, only the data for that particular modifier will be included in the stack-up dimension.

Specifying a Positive or Negative Value

Each dimension in the stack-up must be specified to be either added or subtracted from the computation.

Add: Enter a plus sign (+).
Subtract: Enter a minus sign (-).
System prompt is +/-.
Response is entered following prompt.

Example 3

#n# SEL DIM TOL BOTH 0.002 PREC BOTH 3 TYPE LIMIT ON [CR]

#n# INS LDI : *DRAW/MODEL ent-end* d_1d_2 *DRAW loc* d_3 *DRAW/MODEL
        ent-end* d_4d_5 *DRAW loc* d_6 *DRAW/MODEL ent-end* d_7d_8
        DRAW loc d_9 [CR] (Fig. 2.58)

#n# CALC TSTack:

	Max.	Min.	Norm	Stat. Tol.	Tol.	Real
DRAW/MODEL ent D_{10} +/-+	10.502	10.498	+ 10.500	+/-0.002	0.000003	
DRAW/MODEL ent D_{11} +/--	4.502	4.498	+ 4.500	+/-0.002	0.000003	
DRAW/MODEL ent D_{12} +/--	4.502	4.498	+ 4.500	+/-0.002	0.000003	
DRAW/MODEL ent CR						
	1.506	1.494	1.500	+/-0.006	0.003463	

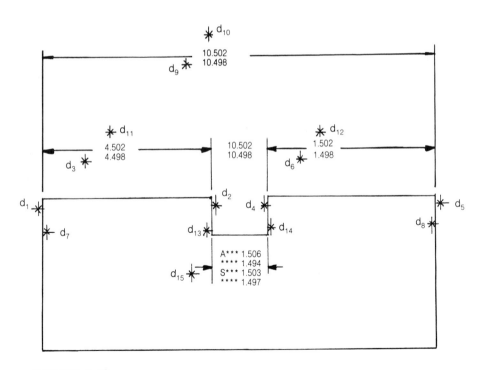

FIGURE 2.58

#n# INS LDI TEXT MAIN /A * * * 1.506 (max. = 1.500 + 0.006)
                      * * * 1.494 (min. = 1.500 - 0.006)
                  S * * * 1.503 (max. = 1.500 + 0.003)
                      * * * 1.497 (min. = 1.500 - 0.003)

/ mod : *DRAW/MODEL ent–end* $d_{13}d_{14}$ *DRAW loc* d_{15} [CR]

2.10 THE THREE-DIMENSIONAL MODEL: PART DEPTH, VIEWS, AND CONSTRUCTION PLANES

2.10.1 Two-Dimensional Model

In working with a two-dimensional model, all entities are entered on a fixed view space coordinate system, (Fig. 2.59) the plane of the screen. In reality, this is the top view, which is the default plane where view space and model space axes are the same.

2.10.2 Three-Dimensional Model

A drawing with views must be provided to see the model. Any number of views can be defined to view the model properly. The system automatically provides seven predefined views on a permanent basis: the six basic third-angle orthographic views plus an isometric view. An infinite number of views can be user defined through construction planes.

Model Space

The three-dimensional model resides in what is referred to as a model space coordinate system (Fig. 2.60). The axes of the part do not change as the part is oriented.

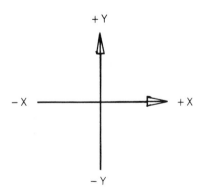

FIGURE 2.59 Two-dimensional coordinate system.

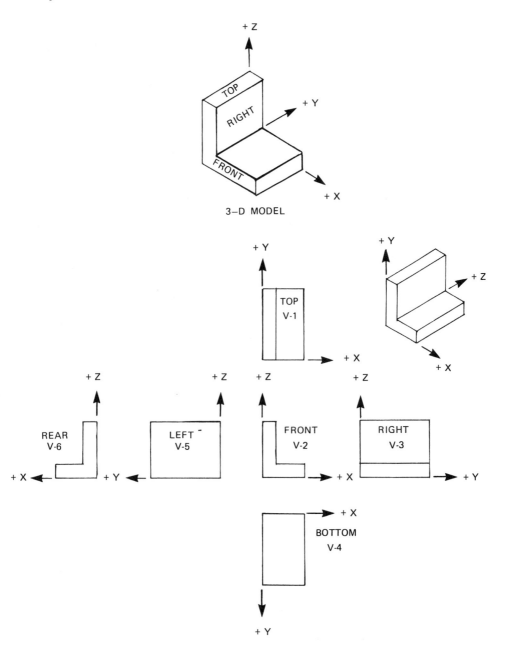

FIGURE 2.60 Three-dimensional model: six basic views plus isometric view, model space axes.

View Space/Construction Plane

Construction or view space is a local coordinate system that is defined by the construction plane for the associated view. The construction space coordinate system may be oriented differently from the model space coordinate system, thus permitting different views of the three-dimensional model. In three dimensions, entities can be entered or projected on the Z axis, which is perpendicular to the X and Y axes (Fig. 2.61).

Preplanning

During the planning process, the user must determine what view or views and which construction plane to work with initially in the construction of the model. This may vary considerably depending on the configuration of the part and the approach taken by the individual user. In the evolution of the part's development, additional views and construction planes can be added and/or deleted as the need arises. The system provides the user with the capability to work in one or several views during the model construction process.

All construction—entering or editing of entities of the model regardless of the active view—is applicable to all views when working in Model mode. As noted previously, views are defined through construction planes. There are seven permanently defined construction planes. An additional infinite number of construction planes and views can be user defined.

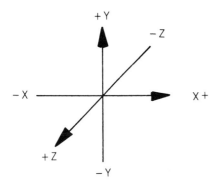

FIGURE 2.61 Three-dimensional coordinates system.

Name	Alternative Names
1	Top, XY, MODEL (default)
2	FRONT, XZ
3	RIGHT, YZ
4	BOTTOM, X-Y
5	LEFT, -YZ
6	REAR, -XZ
7	ISO

2.10.3 View Commands

DEFINE VIEW

Purpose: A view provides a "window" through which part geometry can be seen. Each drawing can contain as many views as needed.

In two dimensions one view was defined and it was relative to construction plane 1, which is the default. In construction plane 1, model space and view space coordinates are the same.

In three dimensions the user determines the number of views that are required and then defines them. The system provides two methods of defining views, explicit reference to a construction plane, and folding. This material covers the construction plane method.

Syntax

#n# DEF VIE viewname CPL cplname (mod) : *DRAW loc* d_1 *DRAW loc*
d_2d_3 [CR]

viewname	Unique name; maximum of 20 characters.
cplname	Name of construction plane to be used in defining view; system or user defined.
d_1	View origin, X0Y0Z0.
d_2d_3	Diagonals defining viewing clipping window. (Do not enter d_2d_3 unless clipping is desired.)

Modifiers

SCALEn   There are three methods of expressing scale of view:

SCALE 2.5	Single numeric value.
SCALE 5 to 2	Ratio of undimensioned numbers.
SCALE 10 in to 1 KM	Ratio of different units.
PERSP	Establishes a perspective view with X and and Y coordinates of the vantage point set to the origin of the view, and the 2 coordinate set at 64 cm.

Example 1

A part (default is three dimensions) and drawing have been activated (default is Model mode), and the drawing frame has been turned on. The user has determined that the three basic views (top, front, right) and the Isometric are best suited for initially defining this particular part.

a. #n# DEF VIE TOP CPL TOP : *DRAW loc* d_1 *DRAW loc* $d_2 d_3$ [CR] (Fig. 2.62)

Explanation

View named TOP defined through construction plane named TOP, added in upper left-hand corner of the screen. Alternative names of construction plane can be used.

b. #n# DEF VIE FRONT CPL FRONT : *DRAW loc* d_1 *DRAW loc* $d_2 d_3$ [CR] (Fig. 2.63)

Explanation

View named FRONT defined through construction plane named FRONT, added in lower left-hand corner of the screen.

c. #n# DEF VIE ISO CPL 7 : *DRAW loc* d_1 *DRAW loc* $d_2 d_3$ [CR] (Fig. 2.64)

Explanation

View named ISO defined through construction plane named 7, added to upper right-hand corner of screen. Note alternative construction plane named 7 used.

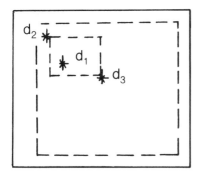

FIGURE 2.62

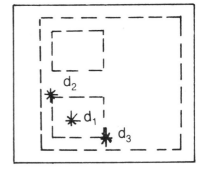

FIGURE 2.63

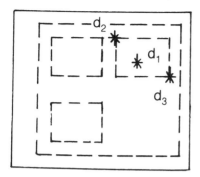

FIGURE 2.64

d. #n# DEF VIE RIGHT CPL YZ : *DRAW loc* d_1 *DRAW loc* d_2d_3 [CR]
(Fig. 2.65)

Explanation

View named RIGHT defined through construction plane named YZ, added to lower right-hand corner of screen. Note alternative construction plane named YZ used.

Example 2

As an alternative, the user may have determined that it was preferable to work in the isometric view only.

#n# DEF VIE ISO CPL ISO : *DRAW loc* d_1 [CR] (Fig. 2.66)

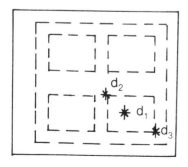

FIGURE 2.65

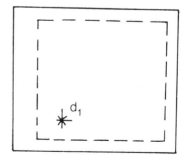

FIGURE 2.66

Explanation

View named ISO defined through construction plane named ISO, which fills the screen. Clipping window is coincident with the drawing extents.

REVISE VIEW

The REVISE VIEW command allows for permanently revising:

1. The size of the clipping window
2. The location of a view

Syntax

1. Revise size of clipping window

   #n# REV VIE CLIP : view d_1 *DRAW loc* d_2d_3 [CR]
      d_1 Identifies view affected.
      d_2d_3 Diagonals of new boundaries.

2. Revise location of view

   #n# REV VIE LOC : view d_1; *DRAW loc* d_2d_3 [CR]
      d_1 Identifies view affected.
      ;   Semicolon as separator.
      d_2 Location is moved to d_3 location.

Example 3

a. Revising clipping boundaries

   #REV VIE CLIP : *view* d_1 *DRAW loc* d_2d_3 [CR] (Fig. 2.67)

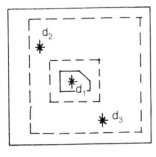

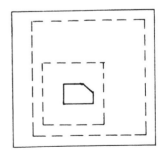

BEFORE   AFTER

FIGURE 2.67

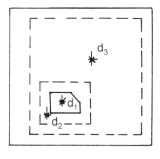

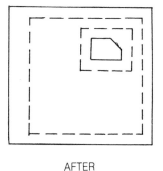

BEFORE   AFTER

FIGURE 2.68

b. Revising view location

#REV VIE LOC : *view* d_1; *DRAW loc* d_2d_3 [CR] (Fig. 2.68)

DELETE VIEW

The DELETE VIEW command will permanently delete the display of one or more views from the currently active drawing. Deleted view names may be reused on newly defined views.

Syntax

#n# DEL VIE : view d_1d_2 · · · [CR]   or   :*view* NAME name of view [CR]
    d_1d_2 · · · View(s) to be deleted is(are) identified by digitizing.

LIST VIEW

Entering the LIST VIEW command provides the user with a listing of status information relative to the active part.

Syntax

#n# LIST VIE (modifiers) [CR]   or   : *view* d_1 [CR]
    d_1 Selects view(s) by digitizing.

Modifiers (Partial Listing)

ALL   Provides selected status for all views. No digitizing required.
SCALE   Scale of view selected.
ORIGIN   Drawing space location of origin of view.
STATUS   Provides all information for view(s) selected. This modifier overrides all other information modifiers.

Example 4

#n# LIST VIEW ALL [CR]

The following views are in drawing Model:

View name: TOP
View name: FRONT
View name: ISO

2.10.4 Construction Planes and Commands

Construction planes (cplanes) are design aids that perform two basic functions:

1. A cplane is either a predefined or user-defined plane which is a local coordinate system on which, when "active" digitizes and explicit input may be entered.
2. Cplanes are used to define the orientation of a view, allowing the model to be displayed from different angles and positions.

DEFINE CPLANE

The DEFINE CPLANE command allows the user to define cplanes for later use with other commands, such as DEFINE VIEW and SELECT CPLANE. Cplanes may have any orientation relative to the model (default) or view coordinate system. Cplanes may be defined by explicit input or digitizing.

Syntax

#n# DEF CPL name (mod) [CR] or
            : *MODEL loc* d_1 [CR] or
            : *MODEL loc* $d_1 d_2 d_3$ [CR]

Name is name of CPL being defined; 20 characters maximum.

Digitizing

None	Orientation specified by modifier.
One digitize	Specifies the origin of the CPL.
Three digitizes	d_1 is origin.
	d_2 is a point on the positive X axis.
	d_3 is a point on the positive Y axis.

Modifiers

FROM name	Name of existing CPL being referenced for new CPL.
AYn	Angle of rotation about Y axis.

AXn	Angle of rotation about X axis.
AZn	Angle of rotation about Z axis. Axes are order dependent. Rotation of AX30 AY25 is not the same as AY25 AX30.
MODEL	Specifies angle of rotation is about model axis. Default.
Viewname	Specifies angle of rotation is about axes of the CPL or digitize of the named or digitized view.

Cplane Axes Rotation

Right-hand rule of thumb (Fig. 2.69): Used to determine positive rotation about axis. Point thumb or right hand in position direction of axis. The direction the fingers point is the positive rotation about that axis. Rotation can be specified in the negative direction by applying the negative or minus (-) sign to the value.

SELECT CPLANE

The SELECT CPLANE command allows the user to select a new currently active cplane. The selected cplane may be one of the seven permanent cplanes provided by the system, or a cplane previously defined by the user with the DEFINE CPLANE command. The selected construction plane becomes active and is used to determine the coordinates of entities inserted with explicit input and the direction of projected entities. The default cplane is No. 1 = TOP = XY = Model and is the active cplane until another one is selected.

Syntax

#n# SEL CPL name [CR]

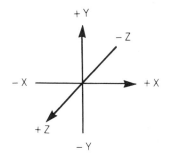

FIGURE 2.69

Name is the name of the defined cplane. There are no modifiers nor any digitizing with this command. A default parameter is being changed. The status display in the upper left-hand corner of the CRT automatically changes to reflect the name of the newly specified active cplane.

ECHO CPLANE

The ECHO CPLANE command allows the user to display temporary graphics of arrows representing the X and Y axes of the current active construction plane (Fig. 2.70). The size of the graphics is user defined, with the origin of the axes at the intersection of the arrows. An "X" appears in the arrowhead on the X axis. The arrows remain visible until turned off. When the cplane arrows are visible, they are automatically updated to represent a newly selected active cplane.

Syntax

#n# ECH CPL (mod) [CR]

Modifiers

LN   Length of arrow is n. Default is n = 2.
OFF   To turn off cplane graphics.

LIST CPLANE

The LIST CPLANE command provides the user with data relative to the construction planes of the active part. The listing may be specified for all cplanes on a cplane name basis.

Syntax

#n# LIS CPL (mod) [CR]

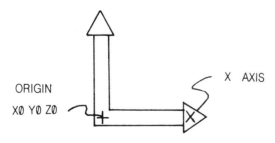

FIGURE 2.70

Modifiers

names	Names of cplanes separated by a space.
ALL	List for all cplanes.
no modifier	List all cplanes, excluding the seven standard ones.

Example 5

(The user has not defined any cplanes.)

#n# LIST CPL ALL [CR]

	Origin in model space			Normal vector in model space		
	OX	OY	OZ	NX	NY	NZ
1 = XY = TOP = MODEL	0.000	0.000	0.000	0.000	0.000	1.000
2 = XZ = FRONT	0.000	0.000	0.000	0.000	-1.000	0.000
	0.000	0.000	0.000	1.000	0.000	0.000
	0.000	0.000	0.000	0.000	0.000	-1.000
	0.000	0.000	0.000	-1.000	0.000	0.000
	0.000	0.000	0.000	0.000	1.000	0.000
	0.000	0.000	0.000	0.577	-0.577	0.577

DELETE PLANE

The DELETE CPLANE command provides for deleting any user-defined construction planes.

The seven standard cplanes cannot be deleted.
The active cplane cannot be deleted.

Syntax

#n# DEL CPL (modifier required) [CR]

Modifiers

ALL	Deletes all user-defined cplanes.
names	Names of cplanes to be deleted separated by a space.

Example 6 (Fig. 2.71)

a. #n# ECH CPL L2 [CR] (Fig. 2.72)

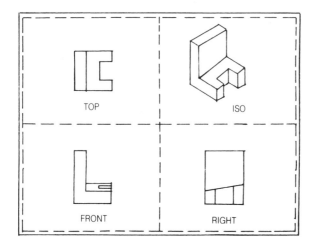

FIGURE 2.71

Explanation

Two-inch-long arrows appear in TOP view. Default active CPL
TOP.

b. #n# SEL CPL FRONT [CR] (Fig. 2.73)

Explanation

CPL arrows automatically appear in FRONT view. CPL FRONT or
2 is now active.

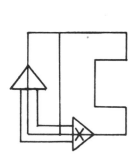

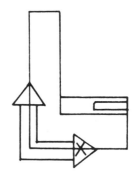

FIGURE 2.72   **FIGURE 2.73**

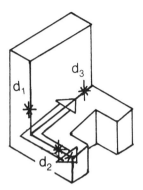

FIGURE 2.74

c. #n# DEF CPL BEVEL : *MODEL loc* END $d_1 d_2 d_3$ [CR] (Fig. 2.74)
   d_1 = origin, d_2 = +X, d_3 = +Y

d. #n# SEL CPL BEVEL [CR] (Fig. 2.74)

Explanation -

User-defined CPL named BEVEL created and made active.

Example 7 (Fig. 4.75)

a. #n# SEL CPL 2 [CR]
   #n# DEF CPL ABC FROM 2 : *MODEL loc* END d_1 [CR]
   #n# SEL CPL ABC [CR]
   #n# ECH CPL L2 [CR] (Fig. 2.76)

Explanation

User-defined CPL named ABC created, referenced from CPL named 2, made active and visible.

b. #n# DEF CPL XYZ FROM ABC AX30 [CR]
   #n# SEL CPL XYZ [CR] (Fig. 2.77)

Explanation

User-defined CPL named XYZ created, referenced from CPL named ABC and 30° in positive direction, made active, arrows automatically updated to active CPL.

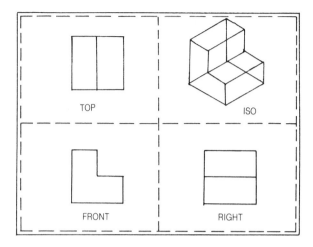

FIGURE 2.75

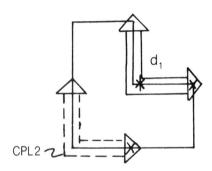

FIGURE 2.76

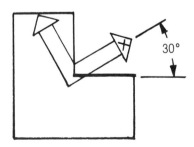

FIGURE 2.77

2.10.5 Project Entity/Part Depth

To define a part in three-dimensions, depth is required. The PROJECT ENTITY command provides the ability to project previously entered entities a specified distance along the Z axis of the construction plane (Fig. 2.78), creating a three-dimensional part or along a specified vector. The depth selected can be either positive or negative.

  Positive (+) projects out of the screen.
  Negative (-) projects into the screen.

The three-dimensional part is referred to as a "wire frame" or "stick" figure and is comprised of a duplicate set of entities, the specified distance interconnected by lines.

Syntax

Projected along the Z axis.

#n# PROJ ENT Dn : *MODEL ent* $d_1 \cdots d_x$ [CR]
  Dn   Indicates a depth of n.
  D1   Indicates a positive depth of 1 in. out of screen.
  D-1   Indicates a negative depth of 1 in. into screen.
  $d_1 \cdots d_x$ Entities to be projected.

Example 7 CPLANE TOP is active.

#n# PROJ ENT D4 : *MODEL ent* CHN d_1 [CR] (Fig. 2.79)

The part is now represented by two rectanges 4 in. apart, interconnected by lines. To the experienced drafter they will appear as surfaces, but they are not. Surfaces are covered later. The six basic

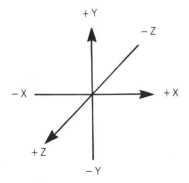

FIGURE 2.78 Model axes coordinates.

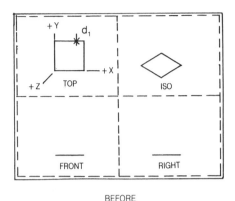

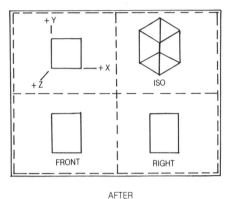

BEFORE   AFTER

FIGURE 2.79

orthographic views and the isometric view, together with user-defined
views, may now be defined with the DEFINE VIEW and DEFINE
CPLANE commands.

Syntax

Projected along vector.

#n# PROJ ENT VEC : *Model ent* d_1 $\cdots$ d_n; *MODEL ent* d_xd_y [CR]
   d_1 $\cdots$ d_n Entities to be projected.
   d_xd_y With the modifier vector, a semicolon is entered
          following identifying the entities to be projected
          and the system responds with the *MODEL end*
          prompt. The user must enter two endpoints to
          define the vector. When the modifier Dn is *used*
          to specify distances of projection, the vector
          indicates direction only of the projection. When
          the modifier Dn is *not used*, the vector defines
          the direction and magnitude of the projection.
          The *end* prompt may be replaced by the user with
          any of the other acceptable prompts: LOC, ORG,
          etc.

Example 8

a. #n# PROJ ENT D4 VEC : *MODEL ent* d_1; *MODEL ent* d_2 LOC IYlIXI
   [CR] (Fig. 2.80)

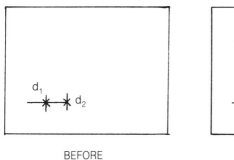
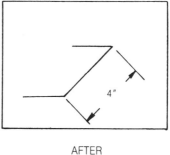

BEFORE   AFTER

FIGURE 2.80

Explanation

Magnitude per D4 is 4 units. Direction per vector of IX1IY1 from d.

b. #n# PROJ ENT VEC : *MODEL ent* d$_1$; *MODEL end* d$_2$d$_3$ [CR] (Fig. 2.81)

Explanation

Magnitude and direction are both specified by digitizing the ends of an existing vector (line).

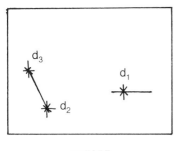
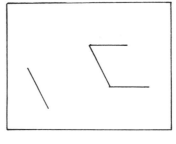

BEFORE   AFTER

FIGURE 2.81

2.11 WORKING WITH VIEWS: THE THREE-DIMENSIONAL MODEL

2.11.1 Editing a Multiview Drawing

There are applications where the user may want to clarify or enhance a multiview drawing. The system for clarifying or enhancing a sel-elected view or views of the model is through editing. This may include one or more of the following:

1. Adding an entity to one view only.
2. Erasing an existing model entity, which makes the entity invisible in that view only.
3. Changing the line font of entities in one or more selected views.

Add Entity to One View Only

As noted in Section 2.10, when working on the three-dimensional model while in the Model mode, all entities inserted into one view are automatically entered into all views of the part. That is because they are model entities.

Example 1: Entering Model Mode Entity

#n# SEL MODE MODEL [CR]
#n# INS CIR DIAM 2 : *MODEL loc* d_1 [CR] (Fig. 2.82)

Explanation

Circle entered in Model mode appears in all views.

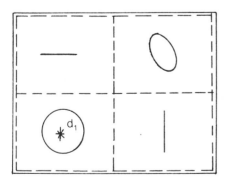

FIGURE 2.82

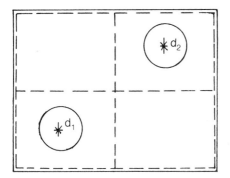

FIGURE 2.83

Example 2: Entering Draw Mode Entity

#n# SEL MODE DRAW [CR]
#n# INS CIR DIAM 2 : *DRAW loc* d_1d_2 [CR] (Fig. 2.83)

Explanation

Circle entered in Draw mode appears in specified view only.

Draw mode only. The basic geometric entities (line, circle, fillet, arc, etc.) can be added to a selected view as Draw mode entities. The system must be in Draw mode. These are Draw mode entities rather than Model entities and must be treated accordingly. Although they may appear to be entered relative to the model, they are entered parallel to the screen. *Construction plane coordinates do not apply to Draw entities.*

ERASE ENTITY and RE-ECHO ENTITY

Model mode only. The ERASE ENTITY command allows the user to erase (make invisible) model entities from a selected view or views. The entity remains in all other views and in the data base. (The DELETE ENTITY command permanently removes an entity from the data base.)

Model or Draw mode. The RE-ECHO ENTITY command allows the users to redisplay previously erased model entities.

Syntax

#n# ERA ENT : *MODEL ent* $d_1 \cdots d_n$ [CR]
There are no modifiers used with this command.

$d_1 \cdots d_n$ Identify entities to be erased.

#n# REE ENT *view* $d_1 \cdots d_n$; *MODEL ent* $d_2 \cdots d_n$ [CR]

#n# REE ENT ALLVIEWS : *MODEL ent* $d_2 \cdots d_n$ [CR]

$d_1 \cdots d_n$ Identify view(s) affected.

;   Semicolon, to switch from *view* to *ent* prompt.

$d_2 \cdots d_n$ Identify model entities to be made visible by digitizing them in any view where they are currently displayed.

Example 3:   Erasing Model Entities

#n# ERA ENT : *MODEL ent* $d_1d_2d_3$ [CR] (Fig. 2.84)

Explanation

Entities identified are made invisible in view 7 only.

Example 4:   Redisplaying Previously Erased Model Entities

#n# REE ENT : view d_1 : *MODEL ent* $d_2d_3d_4$ [CR] (Fig. 2.85)

Explanation

Previously erased model entities are made visible in view 7, although they are identified in a different view.

Change Line Font in View(s)

The line font of an existing model entity can be changed in one or more views with the CHANGE APPEARANCE command. The font of the entity will be affected only in the view or views where it is digitized. See also Section 2.4.1.

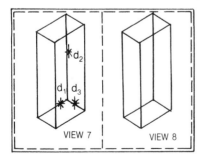

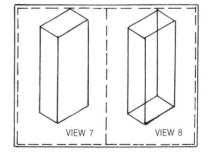

BEFORE   AFTER

FIGURE 2.84

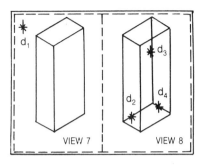 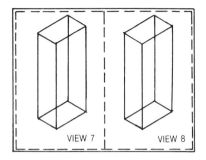

BEFORE   AFTER

FIGURE 2.85

Syntax

#n# CHA APP FONT DASH : *MODEL* ent $d_1 \cdots d_n$ [CR]
$\quad d_1 \cdots d_n$ Identify entities in vews where they will be changed
$\qquad$ to a dash font.

Example 5: Changing Line Font

#n# CHA APP FONT DASH : *MODEL* ent $d_1 d_2 d_3$ [CR] (Fig. 2.86)

Explanation

Entities are changed to dash font only in the view where they are
identified.

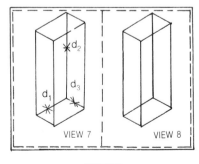 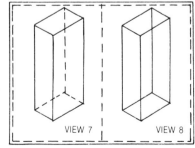

BEFORE   AFTER

FIGURE 2.86

Trim and Divide in Three Dimensions

The INTERSECTION OF modifier applies when working with three-dimensional model entities that appear to intersect in a view although they are at different depths.

Example 6: Trim

#n# TRI ENT INTOF : *MODEL ent* d_1 *MODEL ent* d_2 [CR] (Fig. 2.87)

Explanation

Even though lines AB and CD appear to intersect in the front view, they are at different depths, as shown in the right view. Line AB is trimmed to the apparent point of intersection with CD in the front view.

Example 7: Divide

#n# DIV ENT INTOF : *MODEL ent* d_1 *MODEL ent* d_2 : *MODEL ent* d_3
MODEL ent d_4 [CR] (Fig. 2.88)

Explanation

Lines AB and BC are both divided into two separate lines at the apparent point of intersection in the three-dimensional view.
The view can now be further enhanced to clarfiy the part by:

1. Changing lines to dash font to represent hidden features

#n# CAH APP FONT DASH : *MODEL ent* d_1d_2 [CR] (Fig. 2.89)

2. Making the lines invisible

#n# ERA ENT : *MODEL ent* d_1d_2 [CR] (Fig. 2.90)

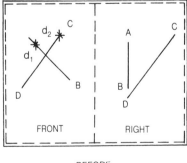 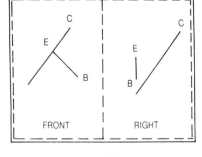

BEFORE   AFTER

FIGURE 2.87

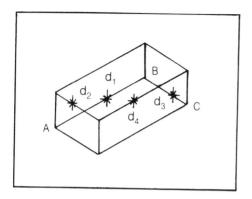

FIGURE 2.88

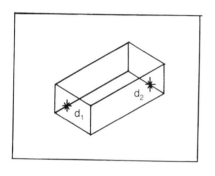

BEFORE

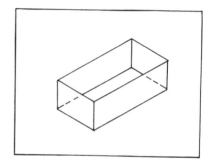

AFTER

FIGURE 2.89

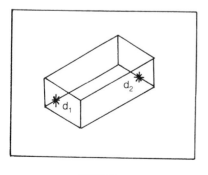

BEFORE

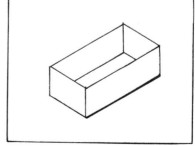

AFTER

FIGURE 2.90

Working in the Isometric View

Entities can be entered explicitly relative to the active construction plane or relative to existing entities.

Example 8

a. #n# DEF VIE 7 CPL 7 : *view loc* d_1 [CR]
    System default is CPL 1.

    #n# INS LIN : *MODEL loc* X0, X4, Y4, X0, Y0 [CR] (Fig. 2.91)
    Lines define base of model.

b. #n# INS LIN : *MODEL loc* END d_1 LOC Z4:
    MODEL loc END d_2 LOC Z4:
    MODEL loc END d_3 LOC Z1:
    MODEL loc END d_4 LOC Z1 [CR] (Fig. 2.92)

c. #n# INS LIN PAR LNG 2: *MODEL ent* d_1
    MODEL loc END d_2d_3: *MODEL ent* d_4
    MODEL loc END d_5d_6 [CR]

    #n# INS LIN : *MODEL loc* END d_7d_8:
    MODEL loc END $d_9d_{10}d_{11}d_{12}d_{13}$ [CR] (Fig. 2.93)

Example 9: Alternative Method

The base created in Example 8 (Fig. 2.91) is projected and then edited.

a. #n# PROJ ENT D4 : *MODEL ent* CHN d_1 [CR]
    #n# DEL ENT : *MODEL ent* d_2 [CR]
    #n# TRI ENT LNG 2 : *MODEL ent* d_3d_4.
    LNG 1 : *MODEL ent* d_5d_6 [CR] (Fig. 2.94)

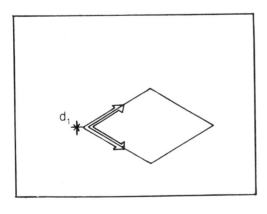

FIGURE 2.91

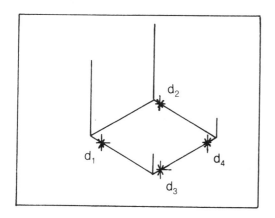

FIGURE 2.92

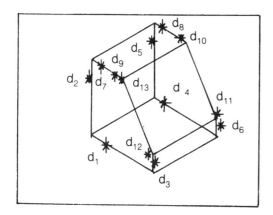

FIGURE 2.93

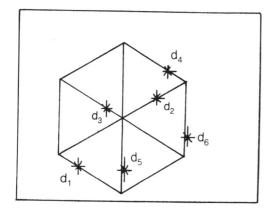

FIGURE 2.94

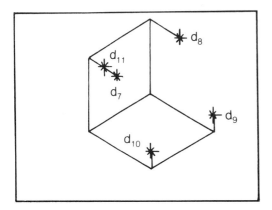

FIGURE 2.95

b. #n# INS LIN : *MODEL loc* END $d_7d_8d_9d_{10}d_{11}$ [CR] (Fig. 2.95) and (2.96)

Example 10

A new construction plane is created to add entities on the sloped surface and the model is further enhanced by making invisible the lines that would be hidden in this view.

#DEF CPL 8 : *MODEL loc* END $d_1d_2d_3$ [CR]
#SEL CPL 8 [CR]
#ECH CPL L2 [CR]
#INS CIR DIAM2 : *MODEL loc* X2Y2 [CR]
#ERS ENT : *MODEL ent* $d_4d_5d_6$ [CR] (Fig. 2.97)

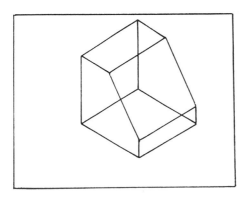

FIGURE 2.96

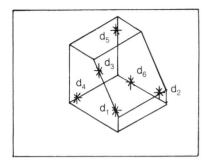 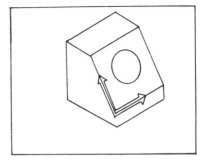

FIGURE 2.97

Example 11

a. A new view is defined relative to construction plane 8, bringing the sloping surface parallel to the screen, creating a true-size view.

#n# DEF CIE 8 CPL 8 :
    DRAW loc d_1 *DRAW loc* [CR] (Fig. 2.98)

b. To enhance the auxiliary view, lines not relative to the surface are made invisible in view 8 only.

#ERA ENT : *MODEL ent* d_1 $\cdots$ d_7 [CR] (Fig. 2.99)

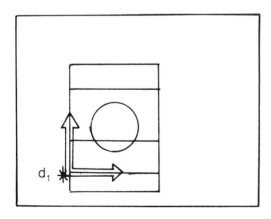

FIGURE 2.98

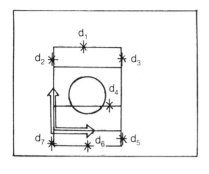

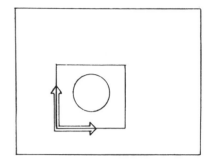

BEFORE   AFTER

FIGURE 2.99

2.11.2 Dimensioning the Isometric View

Linear and angular dimensioning in the isometric view is similar to dimensioning an orthographic view, with the exception that there are additional modifiers. The system does not provide for diameter or radial dimensioning in the isometric view. However, a note with a leader to these entities can be entered with the INSERT LABEL command.

Insert Linear Isometric Dimensions

Linear dimensioning in the isometric view is done in model space— the X, Y, Z axes of the part. The modifiers indicate that the view is isometric and the direction of the dimension extention (witness) lines are relative to the axes of the part. These modifiers are in addition to those introduced in Section 2.9. Dimensions are draw entities and are valid in both Model and Draw modes.

MODEL/DRAW end-ent

Syntax

#n# INS LDI ISO (mod) : *MODEL/DRAW end-ent* d_1d_2 *DRAW loc* d_3
    [CR]
    d_1d_2 Identify ends of entities from which extension lines
                will originate.
    d_3 Location of dimension.

Modifiers

ISO   Required to indicate isometric view to be dimensioned.
  XAX   Direction of extension lines on X axis.
  YAX   Direction of extension lines on Y axis.

ZAX   Direction of extension lines on Z axis.

ABS   "Absolute" means that the extension line direction is parallel to the isometric view and perpendicular to the distance being dimensioned.

Insert Angular Isometric Dimensions

To create an angular dimension requires the additional modifier ISO with the basic command. Other appplicable modifiers relative to parameters still apply.

Syntax

#n# INS ADI ISO mod : *MODEL /DRAW* end d_1d_2 *DRAW loc* d_3 [CR]
    (Fig. 2.100)

    d_1d_2 Identify ends of entities from which extension lines will originate.

    d_3 Location of dimension.

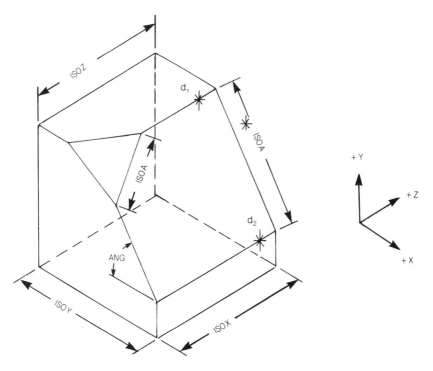

FIGURE 2.100

INSERT LABEL

The INSERT LABEL command allows the user to add a note (text) with a leader to a specified entity to a view of a drawing. A label is a draw entity and is valid in both Model and Draw modes.

Syntax

#n# INS LAB "text" (mod) : *MODEL ent* d_1 *DRAW loc* d_2 [CR]
"text"   Beginning and ending dilimiter and text.
d_1 Identifies entity to which arrow will be drawn.
d_2 Where text will start.

Modifiers

LEA END   Specifies that the leaser will be drawn to the end of
             the text. Default is head.
HEIn   Height of text.

Example 12

#INS LAB "SEE NOTE #1" : *MODEL ent* d_1 *DRAW loc* d_2 [CR] (Fig.
2.101)

2.11.3 Display Control of Views

Section 2.5 DRAWING introduced the user to the display control of drawings with the ZOOM and SCROLL DRAWING commands, which affect the entire drawing, and while they applied to two-dimensional drawings, they still apply to three-dimensional multiview drawings. However, the ZOOM, SCROLL, and DYNAMIC VIEW commands are applicable to three-dimensional multiview drawings only.

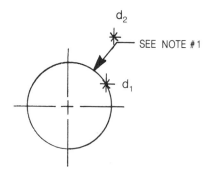

FIGURE 2.101

Whereas the drawing display commands affect all entities on the screen, the view display commands affect only the model entities. Drawing entities associated with a view are not affected. These commands do not permanently alter the drawing. It is a temporary change of the display of a view within a drawing. Views may be returned to their permanent size and orientation with the RESET command.

ZOOM VIEW

Syntax

#n# ZOO VIE WIN : *view* d_1; *MODEL loc* d_2d_3 [CR]
    d_1 Identifies view affected.
    ;   Semicolon to switch to location prompt.
    d_2d_3 Extents of window to be zoomed within view.

or

#n# ZOO VIE (mod) : view $d_1 \cdots d_n$ [CR]
    $d_1 \cdots d_n$ Identifies view(s) affected.

Modifiers

A modifier is required with this command.

IN	The view is increased by twice its present size.
OUT	The view is reduced by one-half its present size.
Rn	The view is redisplayed by a factor of n from its present size.
An	The view is redisplayed by a factor of n to its true size.
ALL	The view is both zoomed and scrolled to just within the clipping window and may appear quite small.
WIN	A window defined by two diagonal digitizes is zoomed and scrolled to fill the clipping window.

SCROLL VIEW

Syntax

#n# SCR VIE : *view* d_1; *MODEL loc* d_2d_3 [CR]
    d_1 Identifies view affected.
    ;   Semicolon to switch to location mode.
    d_2d_3 Scrolling displacement from d_2 to d_3 within clipping window of view.

DYNAMIC VIEW

The DYNAMIC VIEW command requires an Instaview screen. The DYNAMIC VIEW command allows the user to use the image control unit

(ICU) to manipulate the graphic display model entities (zoom scroll, rotate) in the view(s) selected. Up to two views can be acted on at one time. The drawing entities are not affected and are temporarily blanked out upon activation of the ICU.

Image Control Unit (ICU)

Mode button	Director Buttons	
RESET	No direction. Deactivates ICU and returns system to CADDS prompt.	
ZOOM	Up #1	
	Down #2	
SCROLL	Up #1	
	Down #2	
	Left #3	
	Right #4	
ROTATE	Clockwise about X axis,	#1.
	Counterclockwise about X axis,	#2.
	Clockwise about Y axis,	#3.
	Counterclockwise about Y axis,	#4.
	Clockwise about Z axis,	#1 and #2 simultaneously.
	Counterclockwise about Z axis,	#3 and #4 simultaneously.
RESTART	To reset view prior to manipulating without deactivating ICU.	ZOOM, SCROLL, and ROTATE simultaneously.

To operate the ICU, first select a mode and then a direction.

Syntax

#n# DYN VIE (mod) : *view* d_1 [CR]
Activates the ICU. Red light appears on unit. d_1 Identifies view affected.

Modifiers

SPEEDn	Speed for all dynamic commands; n = 1 to 10. Default is 5.
ZSPEEDn	Zoom scale speed, incremental percentage increase/decrease.
SSPEEDn	Scroll speed, incremental percentage.
RSPEEDn	Incremental rotation of graphics in degrees.

RESET VIEW

The RESET VIEW command causes selected views to be returned to their last permanent state. A typical application would be after manipulations of a view with the ZOOM or SCROLL commands. Valid in both Model and Draw modes.

Syntax

#n# RESE VIE : *view* $d_1 \cdots d_n$ [CR]

or

#n# RESE VIE ALL [CR]
  $d_1 \cdots d_n$ Identify views affected.
  With all modifier no digitizing is required.

SET VIEW

The SET VIEW command allows the user to update a view(s) to a new permanent state from a temporary state. Valid in both Model and Draw modes.

Syntax

#n# SET VIEW : *view* $d_1 \cdots d_n$ [CR]

or

#n# SET VIE ALL [CR]
  $d_1 \cdots d_n$ Identify views affected.
  With the ALL modifier no digitizing is required.

2.11.4 Crosshatching View

Crosshatching is a draw entity by default and therefore will be entered only in the construction view. Crosshatching can be entered as a model entity that will appear in all views by adding the modifier MODEL to the basic command. Crosshatching is entered at the depth of the surface of the construction plane regardless of the depth of the bounding entities of the surface digitized. A temporary construction plane may be needed to satisfy this requirement. See also Section 2.4.3.

Syntax

#n# INS XHA MODEL : *MODEL ent* $d_1 \cdots d_n$ [CR]
  $d_1 \cdots d_n$ Bounding entities of surface to be crosshatched.

2.11.5 Grids in Views

Whereas a grid selected in the Draw mode affects the entire drawing (screen), a model grid is related to the model through the viewing windows. Additional orientation modifiers are provided which are applicable and valid only in the Model mode. The layer of grid defaults to the active mode upon selection. See also Section 2.4.4.

Orientation Modifiers

The default orientation in model space is the top view. Construction plane 1; the angle of the grid is relative to the model space.

ANx   Angle of rotation about X axis.
AYn   Angle of rotation about Y axis.
AZn   Angle of rotation about Z axis.
VIEW   When the VIEW modifier is used, the angle of rotation is
        about the currently active construction plane.

Example 13

a. #n# SEL GRI [CR]
    #n# ECH GRI (Fig. 2.102)

Explanation

Defaults to top view model space; 1 X 1 rectangular grid.

3. #SEL GRI AX90 [CR]
    #ECH GRI [CR] (Fig. 2.103)

Explanation

The grid was rotated 90° about the X axis and is now parallel to construction plane 2.

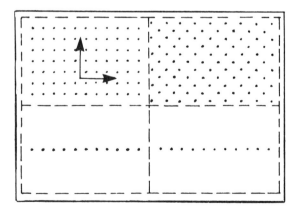

FIGURE 2.102

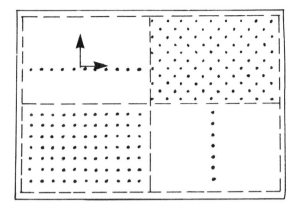

FIGURE 2.103

c. #SEL CPL 2 [CR] (Fig. 2.104)
d. #SEL GRI VIEW AY90 [CR]
   #ECH GRI [CR] (Fig. 2.105)

Explanation

The grid was rotated 90° about the Y axis and is now parallel to construction plane 3.

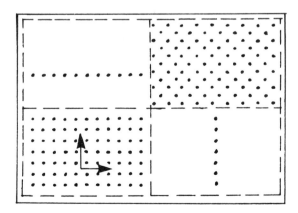

FIGURE 2.104

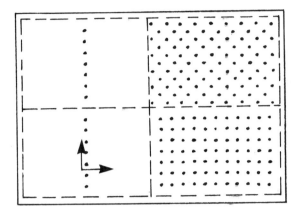

FIGURE 2.105

2.12 SURFACES

2.12.1 Tabulated Cylinders

A tabulated cylinders is a three-dimensional surface generated by the translation of a plane curve along a vector. Two methods are provided for generating tabulated cylinders:

1. The PROJECT ENTITY command with the tabulated cylinder modifier for projecting in one direction only.
2. The INSERT TABULATED CYLINDER command, which allows for projection in two opposite directions.

PROJECT ENTITY

The PROJECT ENTITY command was discussed previously. At this time the modifier TABULATED CYLINDER (TCYL) is introduced. The projection is surfaces rather than a stick figure.

Syntax

#n# PROJ ENT Dn TCYL : *MODEL ent* $d_1 \cdots d_x$ [CR]
$\quad d_1 \cdots d_x$ Existing entities are projected a depth n, generating surfaces.

Example 1: Construction Plane Is Front

#n# PROJ ENT D4 TCYL : *MODEL ent* CHN d_1 [CR] (Fig. 2.106)

Explanation

The four lines projected a depth of 4 have created four surface entities. The original four lines remain as lines.

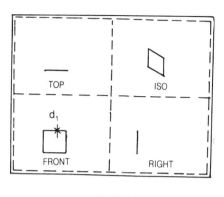 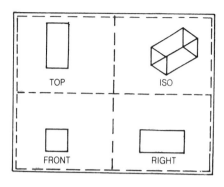

BEFORE   AFTER

FIGURE 2.106

INSERT TABULATED CYLINDER

The INSERT TABULATED CYLINDER command provides for having the beginning and ending boundaries at a different depth than that of the original entities.

Syntax

#n# INS TCYL (mod) : *MODEL* ent $d_1 \cdots d_n$; *MODEL end* $d_x d_y$ [CR]

$d_1 \cdots d_n$ Existing entities are identified which will be projected to generate surface.

; Semicolon as delimiter to get mode prompt.

d_x Reference or starting point.

d_y Defines direction from d_x, the starting point.

The *end* prompt may be changed by the user to any acceptable prompt.

Modifiers

A modifier must be used if the cylinder is to be other than 1 in. long.

LOWBNDn Low boundary. Starting depth of cylinder. Default is 0 in.

HIBNDn High boundary. Ending depth of cylinder. Default is 1 in.

Example 2: Construction Plane Is Top

#n# INS TCYL LOWBND-4 HIBND6 : *MODEL* ent d_1; *MODEL end* d_2 LOC 1X3IY1 [CR] (Fig. 2.107)

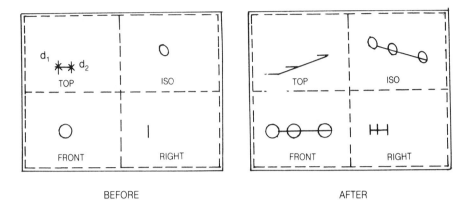

BEFORE   AFTER

FIGURE 2.107

Example 3

To project in or out of the screen, the incremental axis is the Z axis. Construction plane is front. IZ1 = out of screen. IZ-1 = into screen.

#n# INS TCY HIBND 12 : *MODEL ent* d_1; *MODEL end* d_2 LOC IZ1 [CR] (Fig. 2.108)

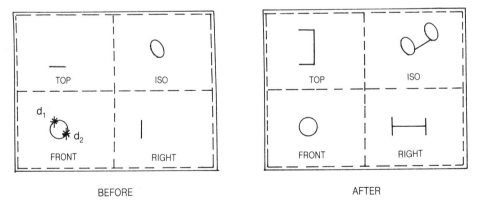

BEFORE   AFTER

FIGURE 2.108

2.12.2 Ruled Surface

A ruled surface is defined by two curves with one end of one curve identified with one end of the other curve. The existing curves (entities) may be lines, circles, arcs, conics, or splines. The ruled surface is defined by straight lines between the two existing curves. Since the ruled surface is an entity in itself, the straight lines that define the boundaries of the surface cannot be edited individually, only as a surface.

Syntax

#n# INS RSUR : *MODEL ent* d_1d_2 [CR]
    d_1d_2 Identify the ends of existing entities between which a surface will be created.

a. #n# INS RSUR : *Model ent* d_1d_2 [CR] (Fig. 2.109)

Explanation

A surface is created between two existing arcs.

b. #n# INS RSUR : *MODEL ent* d_1d_2 [CR] (Fig. 2.11)

Explanation

A line defining a boundary is created between d_1 and d_2.

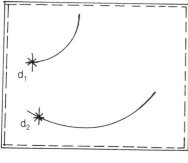
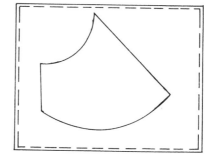

BEFORE   AFTER

FIGURE 2.109

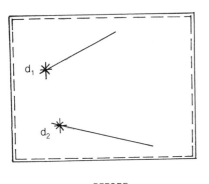
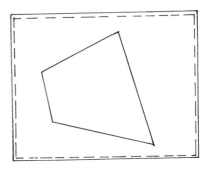

BEFORE   AFTER

FIGURE 2.110

c.   #n# DEL ENT : *MODEL ent* d$_1$ [CR] (Fig. 2.111)

   Explanation

   Nothing will happen because the entity is not a line.

d.   #n# INS RSUR : *MODEL ent* d$_1$d$_2$ [CR] (Fig. 2.112)

   Explanation

   Note the boundary lines between d$_1$ and d$_2$. The shape of the surface is a twist.

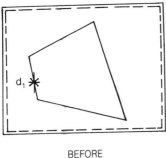
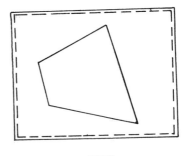

BEFORE   AFTER

FIGURE 2.111

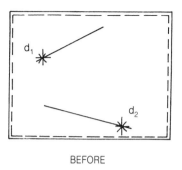
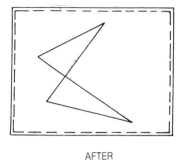

BEFORE   AFTER

FIGURE 2.112

2.12.3 Surface of Revolution

The SURFACE OF REVOLUTION command is used to generate a curved surface of an entity about an axis. The existing entity may be a line, arc, conic, or spline. Each entity is a curve which, when rotated, generates a surface. The surface may be generated at varying degrees about the axis as specified by the modifiers LOWANGLE and HIGHANGLE. The direction from the first position to the second of rotation is determined by the digitizing of the axis of rotation using the right-hand rule of thumb.

Right-hand rule of thumb: Point curved fingers of right hand in the direction of desired rotation; the second digitize on the axis is the direction the thumb is pointing.

Syntax

#n# INS SREV (mod) : *MODEL* ent $d_1 \cdots d_n$; *MODEL* end $d_x d_y$ [CR]
$\quad d_1 \cdots d_n$ Entities to be rotated, generating surfaces.
$\quad$;   Semicolon as delimiter to get mode prompt.
$\quad d_x d_y$ Two digitizes required to define axis of rotation.

The *end* prompt may be changed by the user to any acceptable prompt.

Modifiers

LOWANGn	Low angle. Rotation starts at n degrees. Default is 0°.
HIANGn	High angle. Rotation ends at n degrees. Default is 360°.

Example 5

Three lines rotated 360° (high-angle default) to generate three separate surfaces. Direction of rotation is clockwise as determined by digitizes d_1 and d_2 on axis. Original lines remain at 0! (low-angle default).

#n# INS SREV : *MODEL ent* $d_1d_2d_3$; *MODEL end* d_4d_5 [CR] (Fig. 2.113)

Example 6

The line identified is rotated 180° about the axis in a counterclockwise direction to create a surface.

#n# INS SREV HIANG 180 : *MODEL ent* d_1; *MODEL end* d_2d_3 [CR] (Fig. 2.114)

Example 7

The original line remains. A surface is generated about the axis in a counterclockwise rotation with a low angle of 30° and a high angle of 180°.

#n# INS SREV LOWANG 30 HIANG 180 : *MODEL ent* d_1; *MODEL end* d_2d_3 [CR] (Fig. 2.115)

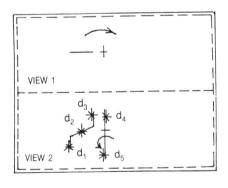

BEFORE

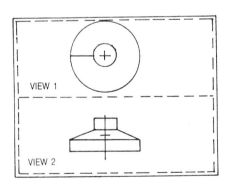

AFTER

FIGURE 2.113

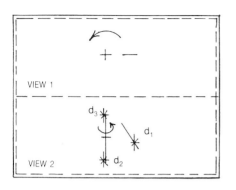
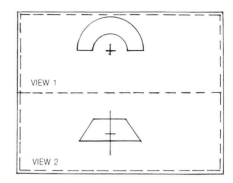

BEFORE   AFTER

FIGURE 2.114

Example 8

Application: To create the same 90° bent tube from each view. The construction plane for the respective view must be active. The circle is rotated about the axis specified by the point and the incremented location to generate the circular 90° form.

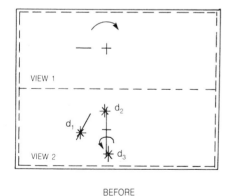
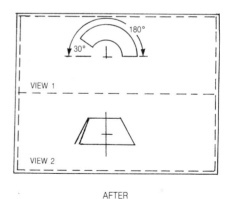

BEFORE   AFTER

FIGURE 2.115

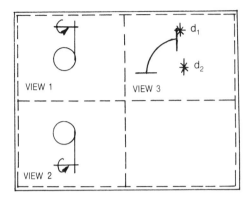

FIGURE 2.116

View 1: #n# INS SREV HIANG 90 : *MODEL ent* d_1; *MODEL end* POI
d_2 LOC IX1 [CR]

View 2: #n# INS SREV HIANG 90: *MODEL ent* d_1; *MODEL end* POI
d_2 LOC IX1 [CR]

View 3: #n# INS SREV HIANG 90: *MODEL ent* d_1; *MODEL end* POI
d_2 LOC IZ1 [CR] (Fig. 2.116)

2.12.4 B-splines and B-surfaces

Although this unit is on surfaces, prior to covering B-surfaces, which
is a surface between two or more existing B-spline curves, a review of
B-splines is included.

B-SPLINE

The B-SPLINE command provides for creating a curve through speci-
fically digitized points. It is equivalent to the drafter's French curve.
There are several interrelated commands with modifiers for creating
and editing B-spline curves which we will not discuss.

This material introduces the user to the B-spline and selected modifiers with the INSERT B-SPLINE command. The command is valid in
both Model and Draw modes.

Syntax

#n# INS BSP (mod) : *MODEL loc* $d_1d_2 \cdots d_n$ [CR]

$d_1d_2 \cdots d_n$ Specific points though which curve is generated.

Modifiers

ZTn   To specify depth at which curve is inserted. Default is
0 in.

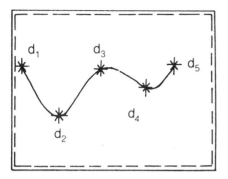

FIGURE 2.117

CIRA   Generates a circular arc through the first three points
        or beginning of B-spline.

CIRB   Generates a circular arc through the last three points
        or ending of B-spline.

Example 9

#n# INS BSP : *MODEL loc* $d_1 d_2 d_3 d_4 d_5$ [CR] (Fig. 2.117)

B-surface

With the INSERT B-SURFACE command, a three-dimensional surface can be created between two or more existing B-spline curves.
The system provides for adjustment of incompatible curves.

Syntax

#n# INS BSURF (mod) : *MODEL ent* $d_1 \cdots d_n$ [CR]
    $d_1 \cdots d_n$ Identify existing B-splines in order of intended
                interpolation.

Modifiers

LAYn   Specifies the layer the surface will be inserted on. De-
        fault is the current layer.

AMIN   Adjust minimum. Adjusts all specified B-splines to the
        B-spline with the least number of defining points.

AMAX   Adjust maximum. Adjusts all specified B-splines to the
        B-spline with the greatest number of defining points.

AINT   Adjust initial. Adjusts all specified B-splines to the
        number of defining points in the initially selected B-
        spline.

PATCH   Displays the patches between the defining points on the
        B-surface.

The following example shows the results of creating a B-surface on three existing B-splines with different numbers of defining points and at different depths. (Fig. 2.118)

   A: five defining points
   B: six defining points
   C: four defining points

   Example 10

a. #n# INS BSURF AMIN PATCH : *MODEL ent* $d_1d_2d_3$ [CR] (Fig. 2.119)

   Adjusting
   Adjustment completed
   Do you wish to continue with interpolation?
   Type Yes to continue; No to exit.
   YES
   Interpolating

b. #n# INS BSURF AMAX PATCH : *MODEL ent* $d_1d_2d_3$ [CR] (Fig. 2.120)

c. #n# INS BSURF AINT PATCH : *MODEL ent* $d_1d_2d_3$ [CR] (Fig. 2.121)

Note: In the examples above the original B-splines have been deleted with the DELETE ENTITY command.

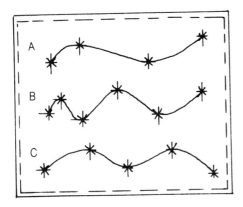

FIGURE 2.118

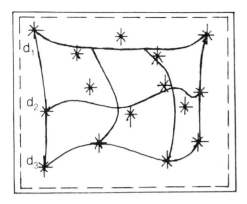

FIGURE 2.119

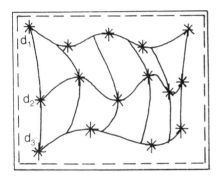

FIGURE 2.120

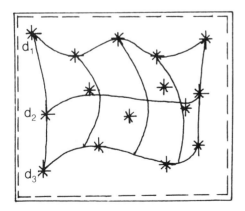

FIGURE 2.121

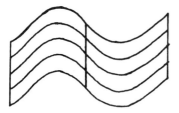

FIGURE 2.122   **FIGURE 2.123**

2.13 SURFACE GRAPHICS

2.13.1 Terminology

Mesh (Fig. 2.122)

Curves displayed on a surface to enhance the visualization.

MESH 2 X 4

Two patches in the interpolating direction (one curve). Four patches in the defining direction (three curves).

Defining Direction (Fig. 2.123)

The mesh curves are the same direction as the initial curve.

Interpolating Direction (Fig. 2.124)

The mesh curves are in the opposite direction of the initial curve.

Patches (Fig. 1.125)

The addition of mesh curves divides the surface into sections referred to as patches.

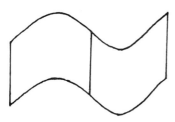

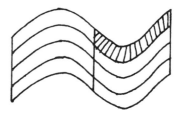

FIGURE 1.124   **FIGURE 2.125**

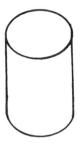

FIGURE 2.126

Hidden Edges (Fig. 2.126)

Edges which are behind a surface and therefore cannot be seen.

2.13.2 Select Surface Graphics

The SELECT SURFACE GRAPHICS command establishes default parameters for surface graphics that will automatically be displayed on all or specified newly created surfaces. Previously created surfaces are not affected.

Syntax

#n# SEL SGR (mod) [CR]

Entity Modifiers

The entity modifiers specify that the appearance modifier specified will apply to all new surfaces of that entity type only.

ALL	All types of surfaces. Default.
TCYL	Tabulated cylinders only.
RSUr	Ruled surfaces only.
SREc	Surfaces of revolution only.
BSUr	B-spline surfaces only.

Appearance Modifiers

MESH nXm	Specifies that the mesh on surfaces will divide each surface into patches n X m.
HED	Hide edge. Hidden edges (boundaries) of cylinders are invisible.
VED	Visible edge. All edges (boundaries) of cylinders are visible.

The edge modifiers are applicable only to 360° cylinders.

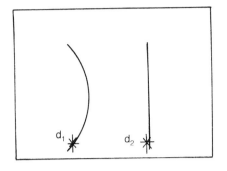

BEFORE

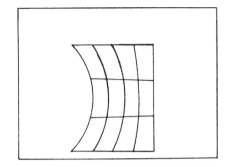

AFTER

FIGURE 2.127

Example 1

Default modifier ALL for entity type is in affect. Surface graphics parameter established so that when surface is created, mesh appears automatically.

a.   #n# SEL SGR MESh 3 X 4 [CR]
    #n# INS RSU : MODEL ent d_1d_2 [CR] (Fig. 2.127)

b.   #n# SEL SGR TCYL VED [CR]
    #n# PROj ENT D6 TCYL : *MODEL ent* d_1 [CR]
    #n# INS SREv LOWANG 0 HIANG 360 : *MODEL ent* d_2; *MODEL loc*
    END d_3d_4 [CR] (Fig. 2.128)

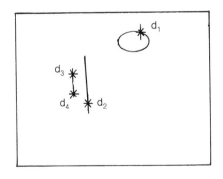

BEFORE

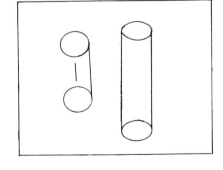

AFTER

FIGURE 2.128

2.13.3 Surface Graphics Specified as Modifier

During the initial creation of a surface, modifiers relative to surface graphics may be entered to:

1. Display a mesh
2. Display or hide edges on cylinders

If surface graphics parameters have previously been established with the SELECT SURFACE GRAPHICS command, the new modifier will override the parameter.

Applicable Surfaces:

Tabulated Cylinders	Surface of revolution
Ruled surfaces	B-spline surface

Syntax

#n# INS (Type of Surface) (Surface Graphics Modifiers) : *MODEL*
    ent $d_1 \cdots d_n$ [CR]

Surface Graphics Modifiers

MESH nXm	Specifies that the mesh on surfaces will divide each surface into patches n x m.
MESH	Mesh without numeric values specifies that preselected mesh rather than edges will be displayed.
HED	Hide edge, Hidden edges (boundaries) of cylinders are invisible.
VED	Visible edge, All edges (boundaries) of cylinders are visible.

The edge modifiers are applicable to only 360° cylinders.

Example 2

A ruled surface is created between a B-spline and a line with a 3 X 4 mesh.

#n# INS RSUr MESH 3 X 4 : *MODEL end* $d_1 d_2$ [CR] (Fig. 2.129)

Example 3

A cylinder is created from a circle with hidden edges invisible.

#n# PROj ENT D10 TCY1 HED : *MODEL ent* d_1 [CR] (Fig. 2.130)

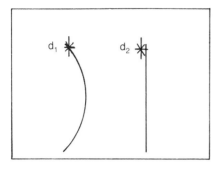
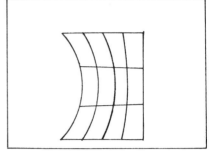

BEFORE   AFTER

FIGURE 2.129

2.13.4 Change Surface Graphics

The CHANGE SURFACE GRAPHICS command allows the user to edit the appearance of existing surfaces.

Syntax

#n# CHA SGR (mod) : *MODEL ent* d_1 [CR]
    d_1 Identifies existing appearance entity to be changed.

Appearance Modifiers

MESH nXm   Specifies that the suface will be changed to a mesh
                 with patches n X m.

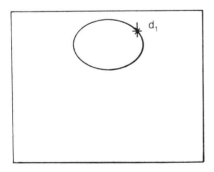
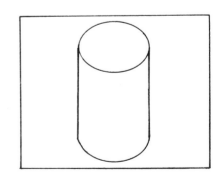

BEFORE   AFTER

FIGURE 2.130

USE MESH	Cylinders only. Without numeric values, specifies that preselected mesh rather than edges will be displayed.
HED	Hide edge. Hidden edges (boundaries) of cylinders become invisible.
VED	Visible edge. All edges (boundaries) of cylinders become visible.
PATch	B-spline surfaces only. Specifies that the mesh will be applied to each B-surface path rather than the surface as a whole. The modofiers MESh 1 X 1 PATch indicate that the B-surface will be displayed with its patch boundaries.

The edge modifiers are applicable only to 360° cylinders.

Example 4

The display is changed from visible to hidden edges.

#n# CHA SGR HED : *MODEL ent* D$_1$ [CR] (Fig. 2.131)

Example 5

The display is changed from 3 X 3 mesh to no mesh (or 1 X 1 mesh).

#n# CHA SGR MESh 1 X 1 : *MODEL ent* d$_1$ [CR] (Fig. 2.132)

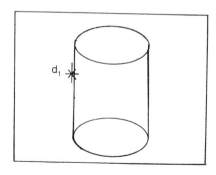

BEFORE

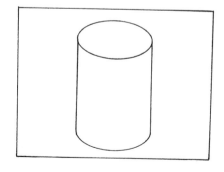

AFTER

FIGURE 2.131

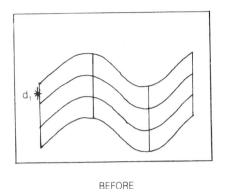

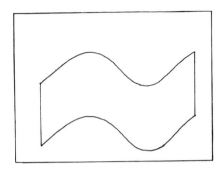

BEFORE   AFTER

FIGURE 2.132

2.14 CUTTING PLANES AND SURFACES

2.14.1 Cut Plane

In mechanical drafting and design, section views are used to clarify the interior of a complex object. A section view is obtained by imagining the object to have been cut by a cutting plane and a portion removed to clearly reveal the interior features. A plane is defined by three noncollinear points. The CUT PLANE command provides three methods of specifying the three points.

1. The endpoints of three existing entities
2. Specifying the plane depth relative to an axis with a modifier
3. Specifying the cutting plane relative to existing entities with location masks

Applicable Entities (Partial List)

1. Curve (plane) entities: When the intersected entity is a curve, points are automatically placed at points of intersection of specified entities and the user is required to add lines between the points to delineate the plane.

Point	Arc	B-spline
Line	Fillet	Conic
Circle	Spline	

2. Surface entities: When the intersected entity is a surface, the entities depicting the plane appear automatically and may be a line, B-spline, arc, or circle.

B-surface	Ruled surface
Tabulated cylinder	Surface of revolution

Note: Surfaces created from a spline entity are not valid.

Mode

Valid in Model mode only.

Basic Syntax

#n# CUT PLAne (mod) : *MODEL end/ent* $d_1 \cdots d_n$ [CR]
    Step One Completed (system response)

Prompt, digitizing, and punctuation vary depending on the modifiers used. See specific syntax for each method of defining cutting planes.

Modifiers

XTn	Specifies the depth of the cutting plane perpendicular
YTN	to either the X, Y, or Z axis at depth n.
ZTN	

NSTPn	Number of steps. Used for specifying multiple cutting planes.

STEPn	Specifies the distance between multiple cutting planes.
	Note: NSTPn and STEPn are used together to specify multiple cutting planes.

`LAYn	Specifies the layer number on which the cutting plane will be located. Default is the current layer.
	Note: If the specified layer is not visible, the cut plane entities will be temporarily displayed until a repaint of the screen. The layer may be made visible with the ECHO LAYER command.

ILAYn	Used with step modifiers to specify incremental layer on which to place successive cutting planes.

CPOINT	Specifies a contact point entity rather than a B-spline when cutting surfaces. Location of points that make up a CPOINT entity can be determined by verification.

Method One

The cutting plane is defined by specifying the endpoints of three existing entities.

Syntax

#n# CUT PLAne (mod) : *MODEL end* $d_1d_2d_3$ *MODEL ent* $d_4 \cdots d_n$
[CR]

$d_1d_2d_3$ Endpoints of three existing entities which are non-
            collinear define the cutting plane.

$d_4 \cdots d_n$ Entities to be cut by the plane.

Example 1

#n# CUT PLA : *MODEL end* $d_1d_2d_3$ *MODEL ent* WIN d_4d_5 [CR]
(Fig. 2.133)

Explanation

All entities are lines. Since the cutting plane is a model entity,
the points of intersection will appear in all views. As this object is
not defined by surfaces, the user must delineate the plane by entering
lines between the points.

Crosshatching the plane. To ensure the crosshatching is inserted
on the plane (correct depth), a construction plane must be defined and
selected on the cutting plane. Crosshatching is by default a draw entity; to crosshatch the plane in all views the modifier MODEL must be
entered.

Method Two

Specifying the plane depth relative to an axis (X,Y, or Z) with a
modifier.

Syntax

#n# CUT PLA ZTn (mod) : *MODEL ent* $d_1 \cdots d_n$ [CR]

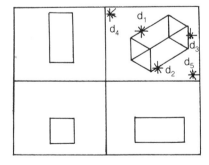

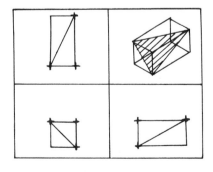

BEFORE   AFTER

FIGURE 2.133

Modifier

ZTn specifies that the cutting plane is perpendicular to the Z axis of the construction plane at a depth of n.

> Prompt is *MODEL ent*
> $d_1 \cdots d_n$ Identify entities to be intersected by plane.

Example 2

#n# CUT PLA ZT6 : *MODEL ent* WIN d_1d_2 [CR] (Fig. 2.134)

Explanation

The active construction plane is Isometric. The object is a tabulated cylinder (surfaces). Since entities are surfaces, the lines delineating the cutting plane will appear automatically.

Method Three

Specifying the cutting plane relative to existing entities with location masks.

This method is similar to Method One except that the *end* mask is canceled by entering any of the other acceptable location masks: ORG, LOC, INTOF, and so on.

Syntax

#n# CUT PLA (mod) : *MODEL end* (mask) $d_1d_2d_3$ *MODEL ent*
    $d_4 \cdots d_n$ [CR]
    Mask   User-entered location mask.

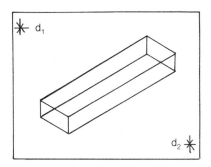

BEFORE

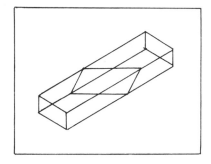

AFTER

FIGURE 2.134

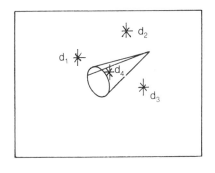

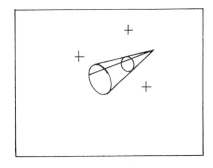

BEFORE   AFTER

FIGURE 2.135

Example 3

#n# CUT PLA : *MODEL end* PO1 $d_1d_2d_3$ *MODEL ent* d_4 [CR] (Fig. 2.135)

Explanation

Object is a surface of revolution (surfaces). *End* locator is canceled and replaced with a point mask. Three points are digitized to define the cutting plane. They are noncollinear and at different depths. The entity depicting the cutting plane is a B-spline.

2.14.2 Cut Surface

In mechanical drafting and design, parts of various geometric shapes are sometimes combined to create a new part. Quite often the adjoining surfaces of these basic shapes meet in intersections that require considerable effort to define. The CUT SURFACE command creates a B-spline or CPOINT entity on the surface specified which represents the intersection of two surfaces. There are three methods of creating the intersecting entity:

1. Both entities are surfaces.
2. Specifying the surface depth relative to an axis with a modifier.
3. A surface intersects a plane.

The two surfaces can be any combination of the following:

Tabulated cylinder	Surface of revolution	Plane
Ruled surface	B-surface	

Mode

Valid in Model mode only.

Basic Syntax

#n# CUT SUR (mod) : *MODEL ent/loc* $d_1 \cdots d_n$
Solution in Progress (system response)
MODEL ent [CR]

Prompt and digitizing varies depending on modifiers used. See specific syntax for each method of cutting surfaces.

Modifiers

XTn	Specifies the depth of the cut surface perpendicular to
YTn	either the X, Y, or Z axis at n.
ZTn	
PLAN	Used when entity to be intersected is a plane. The system responds with the *loc* prompt, requiring the selection of three points lying in the plane.
LAYn	Specifies the layer number on which the cut surface entity will be located. Default is the current layer. *Note:* If the specified layer is not visible, the entity will be temporarily displayed until a repaint of the screen. The layer may be made visible with the *echo layer* command.
CPOINT	Specifies a contact point entity rather than a B-spline.

Method One

Both entities are surfaces.

Syntax

#n# CUT SUR (mod) : *MODEL ent* d_1d_2
Solution in Progress
MODEL ent [CR]
   d_1 Surface to be cut (pierced).
   d_2 Piercing surface.
   Second *MODEL ent* prompt allows for selection of additional surface to be cut.

Example 4

Surfaces are a ruled surface and a tabulated cylinder.

#n# CUT SUR (mod) : *MODEL ent* d_1d_2 [CR] (Fig. 2.136)

Explanation

A B-spline entity is created on the ruled surface (rectangle) where the tabulated cylinder (shaft) intersects or pierces it.

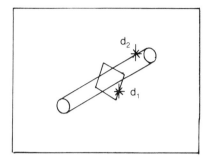

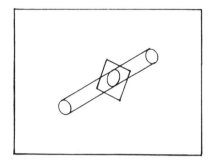

BEFORE   AFTER

FIGURE 2.136

Method Two

Specifying the surface depth relative to an axis with a modifier.

Syntax

#n# CUT SUR ZTn (mod) : *MODEL ent* d_1 [CR]
    The modifier ZTn specifies that the cut surface is perpendicu-
    lar to the Z axis of the construction plane at a depth on n.
    d_1 Only one digitize required to identify surface to be cut.

Example 5

#n# CUT SUR ZT4 : *MODEL ent* d_1

    Solution in Progress
    MODEL ent d_2
    Solution in Progress
    MODEL ent [CR] (Fig. 2.137)

Explanation

    Active construction plane is isometric. Object is two tabulated cy-
linders. Both cylinders were identified through digitizing when second
MODEL ent prompt appeared. Construction plane isometric is active.
A cross section of the tube is defined which is perpendicular to the Z
axis of the active construction plane at a distance of 4 in.

Method Three

A surface intersects a plane.

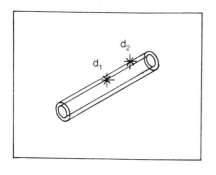
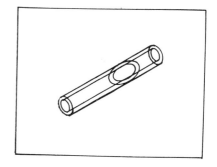

BEFORE   AFTER

FIGURE 2.137

Syntax

#n# CUT SUR PLAN (mod) : *MODEL loc* $d_1d_2d_3$ *MODEL ent* d_4 [CR]
  With the modifier PLAN, the prompt is *loc*, requiring the digit-
  izing of three points to define the plane to be intersected.
  d_4 Piercing surface.

Example 6

The plane is made up of lines and the surface is a tabulated cylinder.

#n# CUT SUR PLAN : *MODEL loc* END $d_1d_2d_3$ *MODEL ent* d_4 [CR]
  (Fig. 1.38)

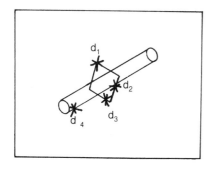
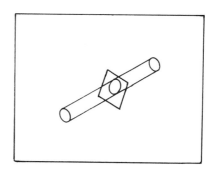

BEFORE   AFTER

FIGURE 2.138

Explanation

The surface to be cut (the plane) is defined by the end mask and the three digitizes. The piercing entity (the shaft) is identified, creating the B-spline entity on the surface.

3
Mass Properties

3.1 BASIC CONCEPTS

The origins of calculus lie in physics and geometry. One branch of physics is concerned with motion, with moving bodies, and with analytical study of the relation between forces applied to bodies and the motion of the bodies under influence of these forces. This branch of physics is called mechanics or dynamics. It is of fundamental importance in the applications of physics to engineering. The concept of motion rests essentially on mathematical notions of space and time. One of the objectives of calculus is to develop the mathematical ideas and tools for understanding and studying motion. Indeed, an exact definition of what is meant by velocity and acceleration is an immediate accompaniment of one of the two main concepts of calculus, that of the derivative of a function.

In studying motion it is essential to develop an understanding of certain aspects of geometry. A moving particle traces out a path, which may be a straight line but which is generally curved. The curve exhibits certain features of the motion, but a full description of the motion requires a correlation between the position of the particle and the time that has elapsed since some initial instant when observation of the motion was begun. It is therefore useful to learn how to investigate curves, how to describe them with algebraic formulas or with formulas of types that transcend algebra, and how to discover their properties in detail by examining the formulas. This kind of thing is part of what is called analytic geometry.

There is another way in which geometry is related to calculus, quite independently of physics and the concept of motion. Geometry is concerned about certain kinds of figures formed by straight lines and planes, and also about certain kinds of curved figures.

Triangles, rectanges, polygons, cubes, prisms, and pyramids are examples of figures formed by straight lines and planes. Circles,
spheres, cylinders, and cones are examples of curved figures. It is a
fundamental matter, in dealing with geometric figures, to know how to
calculate their mass properties, such as circumferences, areas, volumes, first moments, center of gravity, moments of inertia, radius of
gyration, surface area, and so on. In plane geometry the circle is the
simplest curved figures. The circumference of a circle of radius r is
$2\pi r$, and its area is πr^2; here π is a certain number which can be represented approximately by the decimal 3.1416. The precise decimal
representation of π does not terminate, and there is no definite pattern
of repetition in the digits after the decimal points. These measures of
the circumference and area of a circle are arrived at by a method of
limits: the circle is regarded as a limit of inscribed or circumscribed
polygons. There is a much more general method of limits which may be
employed to determine the area of any plane figure which almost completely fills up the inside of the curved figure, but in such a way that
the specially constructed figure is composed entirely of small rectangular pieces, so that its area can be computed simply by adding together
the areas of all these pieces. In order to come closer and closer to
filling up the curved figure, the sizes of the rectangular pieces must
be made smaller and smaller (at least this must be the case of the rectangular pieces near the curved edges); the exact area of the curved
figure is then obtained by a limiting process as shown in Fig. 3.1.

This idea of obtaining the area of a curved figure by a limiting process was used by Archimedes. It is at the root of the concept of the
definite integral of a function. Thus the principal concepts on which
calculus is founded stem, respectively, from the study of motion and
the study of areas of curved figures. That is what was meant when,
at the outset, it was stated that the origins of calculus lie in physics
and geometry.

In this chapter we review the fundamentals of mass properties as
they apply to two- and three-dimensional geometry with examples of

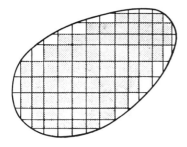

FIGURE 3.1

mass property calculations obtained by using calculus and the interactive computer graphics system. Comparing the results obtained by both methods will enable the reader to understand the accuracy and convenience of an ICGS.

The following mass properties will be investigated:

1. Area
2. Center of gravity for areas
3. Area moments of inertia
4. Polar moment of inertia for areas
5. Radius of gyration for areas
6. Transfer of axes for area moments of inertia
7. Products of inertia for area moments of inertia
8. Transfer of axes for products of inertia
9. Volume
10. Mass
11. Mass moments of inertia
12. Radius of gyration for mass moments of inertia
13. Transfer of axes for mass moments of inertia
14. Products of inertia for mass moments of inertia
15. External surface area

3.1.1 Area

Consider the curve $y = f(x)$ in the interval (a,b) and suppose that the interval is subdivided by the $n + 1$ points $x_1 = a$, x_2, ..., x_i, x_{i+1}, ..., $x_{n+1} = b$ as shown in Fig. 3.2. Erect the corresponding ordinates

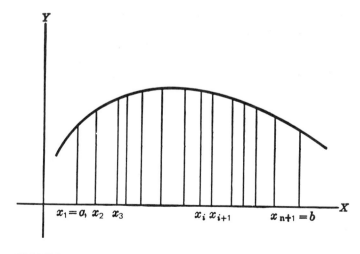

FIGURE 3.2

y_i and write $x_i = x_{i+1} - x_i$ for the width of the ith interval. Now form the sum

$$f(x_1) \, \Delta x_1 + f(x_2) \, \Delta x_2 + \cdots + f(x_n) \, \Delta x_n$$

This sum is usually denoted by the symbol

$$\sum_{i=1}^{n} f(x_i) \, \Delta x_i$$

It seems geometrically evident that this sum approximates the area under the curve between y_1 and y_{n+1} and that in the limit, as the number of points of division becomes infinite and width of each interval approaches zero, it will be equal to the area.

A fundamental theorem of the integral calculus states that

$$\lim_{\substack{n \to \infty \\ \Delta x_i \to 0}} \sum_{i=1}^{n} f(x_i) \, \Delta x_i = \int_{a}^{b} f(x) \, dx$$

This process of *summation*, as it is called, affords a quick and easy way of setting up definite integrals representing areas, lengths of curves, and so on. For areas in rectangular coordinates the reasoning runs something like this: To find the area under the curve $y = f(x)$, a thin rectangular strip is considered, roughly y high and Δx wide. Its area is $y \, \Delta x$. The sum of all such strips would be $\Sigma \, y \, \Delta x$ and the limiting value of the sum of all such strips, $\lim \Sigma \, y \Delta x$, by the fundamental theorem would be equal to $\int_{a}^{b} y \, dx$. This process is shorten by the following reasoning. The area of one strip would be $y \, dx$ and the sum of all such strips would be $\int_{a}^{b} y \, dx$. For the length of a curve it can be said that one little differential arc element is ds in length and the total length sought is the sum of all such elements, $\int ds$. (This is known to be equal to $s = \int ds = \int_{a}^{b} \sqrt{1 + y'^2} \, dx$.)

3.1.2 Center of Gravity of Area and Mass

For a point mass m_1 lying at a distance r_1 from a line L, the first moment of the mass with respect to the line is defined as Fig. 3.3. For n such particles, the sum

$$\text{First moment} = r_1 m_1 + r_2 m_2 + \cdots + r_n m_n$$

$$= \sum r_i m_i$$

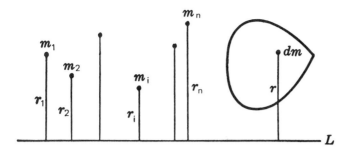

FIGURE 3.3

For a continuous mass distribution this sum becomes an integral:

First moment $= \int r \, dm$

where $\bar{r}$ represents the distance of the element of mass dm from the line L.

The center of mass measured from L is

$$\bar{r} = \text{center of mass} = \frac{\int r \, dm}{\int dm} \tag{1}$$

From Eq. (1) the center of mass can readily be computed for a given mass measured from a given line (or from a given plane in the case where the mass is three-dimensional). Sometimes the center of mass is called the center of gravity (c.g.); for masses that are pure geometrical figures, the term *centroid* is often used.

Consider a plate as an area with a given contour and the mass as density times area, as shown in Fig. 3.4. Thus

$$dm = \rho dA = \rho(y_1 - y_2)dx = \rho(x_1 - x_2)dy$$

and the coordinates of the c.g. are given by

$$\bar{x} = \frac{\int \rho x(y_1 - y_2)dx}{\int \rho(y_1 - y_2)dx}$$

$$\bar{y} = \frac{\int \rho y(x_1 - x_2)dy}{\int \rho(x_1 - x_2)dy} \tag{2}$$

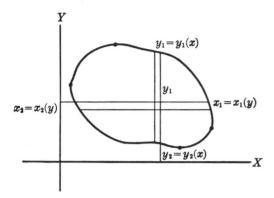

FIGURE 3.4

Note that for $\bar{x}$ the area element is taken as $(y_1 - y_2)\, dx$ since all parts of this strip are at the same distance x from the Y axis; similarly, for $\bar{y}$ all parts of the strip $(x_1 - x_2)\, dy$ are at the same distance y from the X axis. In most problems commonly met, ρ is a constant and in this case it is possible to use the area element $(y_1 - y_2)\, dx$ in computing $\bar{y}$ by considering all the mass of the strip as concentrated at the middle point (i.e., at $y_1 + y_2/2$). When this is done,

$$\bar{y} = \frac{\int (1/2)(y_1^2 - y_2^2)\, dx}{\int (y_1 - y_2)\, dx}$$

and this may be used instead of the expression for $\bar{y}$ in Eq. (2).

But for a three-dimensional mass (volume) the moments are taken with respect to a plane and so define the first moment with respect to that plane. Suppose that the area between the curve $y = f(x)$, $x = 0$, $y = c$, $y = d$, is revolved about the Y axis. Then taking slices perpendicular to the Y axis as was done in computing the volume, the following equation is obtained:

$$dm = \rho \pi x^2\, dy$$

This mass is all at the same distance from the base plane perpendicular to the Y axis. The first moment of this mass with respect to the base plane is shown in Fig. 3.5:

$$\text{first moment} = \pi \int \rho y x^2\, dy$$

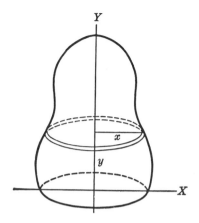

FIGURE 3.5

and the $\bar{y}$ of the c.g. is given by

$$\bar{y} = \frac{\int \rho y x^2 \, dy}{\int \rho x^2 \, dy}$$

Since the solid is one of revolution, this will completely locate the c.g., provided that the solid is homogeneous or provided that the density is a function of the distance from the axis of rotation.

3.1.3 Area Moments of Inertia

When forces are distributed continuously over an area upon which they act, it is often necessary to calculate the moment of these forces about some axis either in or perpendicular to the plane of the area. Frequently, the intensity of the force (pressure or stress) is proportional to the distance of the force from the moment axis. The elemental force acting on an element of area, then, is proportional to distance times differential area, and the elemental moment is proportional to distance squared times differential area. Therefore, the total moment involves an integral of the form $\int (\text{distance})^2 \, d(\text{area})$. This integral is known as the *moment of inertia of the area*. The integral is a function of the geometry of the area and occurs so frequently in the applications of mechanics that it is useful to develop its properties in some detail and to have these properties available for ready use when the integral arises.

Figures 3.6a through 3.6c illustrates the physical origin of these integrals. In Fig. 3.6a the surface area ABCD is subjected to a dis-

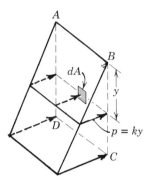

FIGURE 3.6a

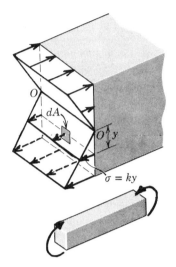

FIGURE 3.6b

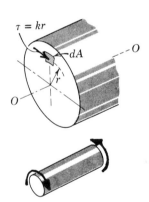

FIGURE 3.6c

tributed pressure p whose intensity is proportional to the distance y from the axis ab. The moment about AB that is due to the pressure on the element of area dA is py dA = ky^2 dA. Thus the integral in question appears when the total moment M = k$\int y^2$ dA is evaluated.

In Fig. 3.6b, the distribution of stress acting on a transverse section of a simple elastic beam bent by equal and opposite couples applied to its ends. At any section of the beam a linear distribution of force intensity or stress σ, given by $\sigma = ky$, is present, the stress being positive (tensile) below the axis O-O and negative (compressive) above the axis. The elemental moment about the axis O-O is $dM = y(\sigma\, dA) = ky^2\, dA$; thus the same integral appears when the total moment $M = k\int y^2\, dA$ is evaluated.

A third example is given in Fig. 3.6c, which shows a circular shaft subjected to a twist or torsional moment. Within the elastic limit of the material this moment is resisted at each cross section of the shaft by a distribution of tangential or shear stress τ, which is proportional to the radial distance r from the center. Thus $\tau = kr$, and the total moment about the central axis is $M = \int r(\tau\, dA) = k\int r^2\, dA$. Here the integral differs from that in the preceding two examples in that the area is normal instead of parallel to the moment axis and in that r is a radial coordinate instead of a rectangular one.

Although the integral illustrated in the preceding examples is generally called the moment of inertia of the area about the axis in question, a more fitting term is the *second moment of area*, since the first moment y dA is multiplied by the moment arm y to obtain the second moment for the element dA. The work "inertia" appears in the terminology by reason of the similarity between the mathematical form of the integrals for second moments of areas and those for the resultant moments of the so-called inertia forces in the case of rotating bodies. The moment of inertia of an area is a purely mathematical property of the area and in itself has no physical significance.

3.1.4 Polar Moment of Inertia for Areas

Consider the area A in the x-y plane (Fig. 3.7). The moments of inertia of the element dA about the x and y axes are, by definition, $dI_x = y^2\, dA$ and $dI_y = x^2\, dA$, respectively. Therefore, the moments of inertia of A about the same axes are

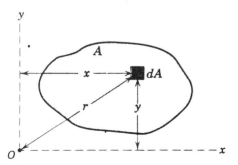

FIGURE 3.7

$$I_x = \int y^2 dA$$
$$I_y = \int x^2 dA \qquad\qquad\qquad (3)$$

where the integration is carried out over the entire area.

The moment of inertia of dA about the pole O (z axis) is, by similar definition, $dJ_z = r^2 \, dA$, and the moment of inertia of the entire area about O is

$$J_z = \int r^2 \, dA \qquad\qquad\qquad (4)$$

The expressions defined by Eqs. (3) are known as *rectangular moments of inertia*, whereas the expression of Eq. (4) is known as the *polar moment of inertia*. Since $x^2 + y^2 = r^2$, it is clear that

$$J_z = I_x + I_y \qquad\qquad\qquad (5)$$

A polar moment of inertia for an area whose boundaries are more simply described in rectangular coordinates than in polar coordinates is easily calculated with the aid of Eq. (5).

The moment of inertia of an element involves the square of the distance from the inertia axis to the element. An element whose coordinate is negative contributes as much to the moment of inertia as does an equal element with a positive coordinate of the same magnitude. Consequently, the area moment of inertia about any axis is always a positive quantity. In contrast, the first moment of the area, which was involved in the computations of centroids, could be either positive, negative, or zero.

3.1.5 Radius of Gyration for Areas

Consider an area A (Fig. 3.8a) which has rectangular moments of inertia I_x and I_y and a polar moment of inertia J_z about O and concentrated into a long narrow strip of area A a distance k_x from the x axis (Fig. 3.8b). By definition the moment of inertia of the strip about the x axis will be the same as that of the original area if $k_x{}^2 A = I_x$. The distance k_x is known as the radius of gyration of the area to be concentrated into a narrow strip parallel to the y axis, as shown in Fig. 3.8c. Also, if the area would be concentrated into a narrow ring of radius k_z as shown in Fig. 3.8d, the polar moment of inertia may be expressed as $k_z{}^2 A = J_z$. In summary, the following equations are written:

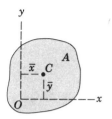

FIGURE 3.8a

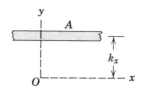

FIGURE 3.8b

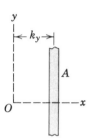

FIGURE 3.8c

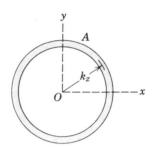

FIGURE 3.8d

$$I_x = k_x^2 A \quad \text{or} \quad k_x = \sqrt{\frac{I_x}{A}}$$

$$I_y = k_y^2 A \qquad\qquad k_y = \sqrt{\frac{I_y}{A}} \qquad\qquad (6)$$

$$J_z = k_z^2 A \qquad\qquad k_z = \sqrt{\frac{J_z}{A}}$$

The radius of gyration, then, is a measure of the distribution of the area from the axis in question. A rectangular or polar moment of inertia may be expressed by specifying the radius of gyration and the area.

When Eqs. (6) are substituted into Eq. (5), the result is

$$k_z^2 = k_x^2 + k_y^2 \qquad\qquad (7)$$

Thus the square of the radius of gyration about a polar axis equals the sum of the squares of the radii of gyration about the two corresponding rectangular axes.

It is imperative that there be no confusion between the coordinate to the centroid C of an area and the radius of gyration. In Fig. 3.8a the square of the centroidal distance from the x axis, for example, is $\bar{y}^2$, which is the square of the mean value of the distances from the elements of the area to the x axis. The quantity k_2^2, on the other hand, is the mean of the squares of these distances. The moment of inertia is not equal to $A\bar{y}^2$, since the square of the mean is less than the mean of the squares.

3.1.6 Transfer of Axes for Area Moments of Inertia

The moment of inertia of an area about a noncentroidal axis may be easily expressed in terms of the moment of inertia about a parallel centroidal axis. In Fig. 3.9 the x_0-y_0 axes pass through the centroid C of the area. Now let's determine the moments of inertia of the area parallel x-y axes. By definition the moment of inertia of the element dA about the x axis is

$$dI_x = (y_0 + d_x)^2 \, dA$$

Expanding and integrating gives

$$I_x = \int y_0^2 \, dA + 2d_x \int y_0 \, dA + d_x^2 \int dA$$

The first integral is by definition the moment of inertia $\bar{I}_x$ about the centroidal x_0 axis. The second integral is zero, since $\int y_0 \, dA = A\bar{y}_0$ and $\bar{y}_0$ is automatically zero with the centroid on the x_0 axis. The third term is simply Ad_x^2. Thus the expression for I_y becomes

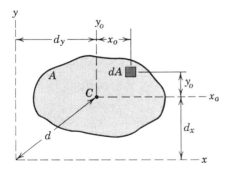

FIGURE 3.9

$$I_x = \bar{I}_x + Ad_x^2$$
$$I_y = \bar{I}_y + Ad_y^2 \tag{8}$$

By Eq. (5) the sum of these two equations gives

$$J_z = \bar{J}_z + Ad^2 \tag{9}$$

Equations (8) and (9) are the *parallel-axis theorems*. Two points in particular should be noted. First, the axes between which the transfer is made must be parallel, and second, one of the axes must pass through the centroid of the area.

If a transfer is desired between two parallel axes neither one of which passes through the centroid, it is first necessary to transfer from one axis to the parallel centroidal axis and then to transfer from the centroidal axis to the second axis.

The parallel-axis theorems also hold for radii of gyration. With substitution of the definition of k into Eqs. (8), the transfer relation becomes

$$k^2 = \bar{k}^2 + d^2 \tag{10}$$

where $\bar{k}$ is the radius of gyration about a centroidal axis parallel to the axis about which k applies and d is the distance between the two axes. The axes may be either in the plane or normal to the plane of the area.

3.1.7 Products of Inertia for Area Moments of Inertia

In certain problems involving unsymmetrical cross sections and in the calculation of moments of inertia about rotated axes, an expression

$$dI_{xy} = xy \, dA$$

occurs which has the integral form

$$I_{xy} = \int xy \, dA \tag{11}$$

where x and y are the coordinates of the element of area $dA = dx \, dy$. The quantity I_{xy} is called the *product of inertia* of the area A with respect to the x-y axes. Unlike moments of inertia, which are always positive of positive areas, the product of inertia may be positive, negative, or zero.

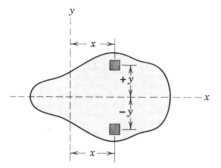

FIGURE 3.10

The product of inertia is zero whenever either one of the reference axes is an axis of symmetry, such as the x axis for the area of Fig. 3.10 and the sum of the terms x(-y) dA and x(+y) dA due to symmetrically placed elements vanishes. Since the entire area may be considered composed of pairs of such elements, it follows that the product of inertia I_{xy} of the entire area is zero.

3.1.8 Transfer of Axes for Products of Inertia

A transfer-of-axis theorem similar to that for moments of inertia also exists for products of inertia. By definition the product of inertia of the area in Fig. 3.10 with respect to the x and y axes in terms of the coordinates x_0, y_0 to the centroidal axes is

$$I_{xy} = \int (x_0 + d_y)(y_0 + d_x) \, dA$$

$$I_{xy} = \int x_0 y_0 dA + d_x \int x_0 \, dA + d_y \int y_0 \, dA + d_x d_y \int dA$$

The first integral is by definition the production of inertia about the centroidal axes, which is noted as $\bar{I}_{xy}$. The middle two integrals are both zero since the first moment of the area about its own centroid is necessarily zero. The third integral is merely $d_x d_y A$. Thus the transfer-of-axis theorem for products of inertia becomes

$$I_{xy} = \bar{I}_{xy} + d_x d_y A \tag{12}$$

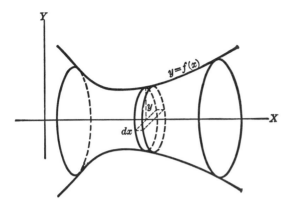

FIGURE 3.11

3.1.9 Volume

Let the area under the curve $y = f(x)$, namely $\int_a^b f(x)\, dx$, be revolved about the X axis, thus generating a volume as shown in Fig. 3.11. The area of a cross section of this solid by a plane perpendicular to the X axis is πy^2. The volume of a little slice dx thick would be $\pi y^2\, dx$. The total volume of the solid of revolution between two parallel planes $x = a$ and $x = b$ would therefore be the sum of all such slices, or

$$V = \pi \int y^2\, dx$$

If the area $\int_c^d x\, dy$ is revolved about the Y axis, the volume becomes

$$V = \pi \int_c^d x^2\, dy$$

Or a cylindrical element of volume can be used. When an elementary area strip as shown in Fig. 3.12 is revolved about the Y axis, the volume element generated is a thin cylindrical shell with area $2\pi x$ $(y_2 - y_1)$. The volume element is $2\pi x(y_2 - y_1)\, dx$; summing these, the volume is obtained:

$$V = 2\pi \int_a^b x(y_2 - y_1)\, dx$$

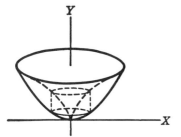

FIGURE 3.12

Consider the sections of a solid made by parallel planes. For purposes of illustration let these planes be perpendicular to the X axis. If it is possible to write down the area of each section in terms of its distance from some fixed point on OX, say the origin O, then the volume of the solid can be determined. For, as in Fig. 3.13, the area of the c cross section at distance x is known to be a function of x, say $\alpha(x)$, and the volume element then is $\alpha(x) \, dx$. Hence the volume from x = a to x = b would be

$$V = \int_a^b \alpha(x) \, dx$$

3.1.10 Mass Moments of Inertia

The equation of moments about an axis normal to the plane of motion for a rigid body in plane motion contains an integral which depends on the distribution of mass with respect to the moment axis. This in-

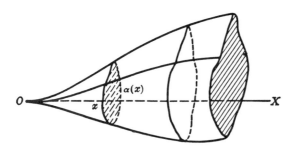

FIGURE 3.13

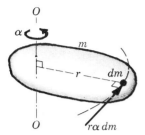

FIGURE 3.14

tegral occurs whenever a rigid body has an angular acceleration about its axis of rotation.

Consider a body of mass m (Fig. 3.14) rotating about an axis O-O with an angular acceleration α. All particles of the body move in parallel planes that are normal to the rotation axis O-O. Any one of the planes can be chosen as the plane of motion, although the one containing the center of mass is usually the one so designated. An element of mass dm has a component of acceleration tangent to its circular path equal to $r\alpha$, and by Newton's second law of motion the resultant tangential force on this element equals $r\alpha$ dm. The moment of this force about the axis O-O is $r^2\alpha$ dm. For a rigid body α is the same for all radial lines in the body and α is taken outside of the integral sign. The remaining integral is known as the moment of inertia of the mass m about the axis O-O and is

$$I = \int r^2 \, dm$$

This integral represents an important property of a body and is involved in the force analysis of any body that has rotational acceleration about a given axis. Just as the mass m of a body is a measure of the resistance to translational acceleration, the moment of inertia is a measure of resistance to rotational acceleration of the body.

The moment-of-inertia integral may be expressed alternatively as

$$I = \sum_i r_i^2 \, m_i$$

where r_i is the radial distance from the inertia axis to the representation particle of mass m_i and where the summation is taken over all particles of the body.

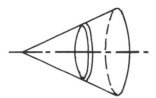
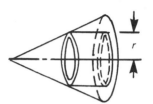

FIGURE 3.15a   FIGURE 3.15b

If the density ρ is constant throughout the body, the moment of inertia becomes

$$I = \rho \int r^2 \, dV$$

where dV is the element of volume. In this case the integral by itself defines a purely geometrical property of the body. When the density is not constant but is expressed as a function of the coordinates of the body, it must be left within the integral sign and its effect accounted for in the integration process.

In general the coordinates that best fit the boundaries of the body should be used in the integration. It is particularly important that a good choice of the element of volume dV be made. An element of lowest possible order should be chosen, and the correct expression for the moment of inertia of the element about the axis involved should be used. For example, in finding the moment of inertia of a solid right circular cone about its central axis, an element can be chosen in the form of a circular slice of infinitesimal thickness, (Fig. 3.15). The differential moment of inertia for this element is the expression for the moment of inertia of a circular cylinder of infinitesimal altitude about its central axis. Alternatively, an element could be chosen in the form of a cylindrical shell of infinitesimal thickness as shown in Fig. 3.15b. Since all of the mass of the element is at the same distance r from the inertia axis, the differential moment of inertia for this element is merely r^2 dm, where dm is the differential mass of the elemental shell.

3.1.11 Radius of Gyration for Mass Moments of Inertia

The radius of gyration k of a mass m about an axis for which the moment of inertia is I is

$$k = \sqrt{\frac{I}{m}} \quad \text{or} \quad I = k^2 m \tag{13}$$

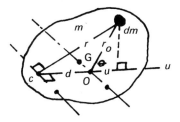

FIGURE 3.16

Thus k is a measure of the distribution of mass of a given body about the axis in question, and its definition is analogous to the definition of the radius of gyration for second mements of area. If all the mass m could be concentrated at a distance k from the axis, the correct moment of inertia would be $k^2 m$. The moment of inertia of a body about a particular axis is frequently indicated by specifying the mass of the body and the radius of gyration of the body about the axis. The moment of inertia is then calculated from Eq. (12).

3.1.12 Transfer of Axes for Mass Moments of Inertia

If the moment of inertia of a body is known about a centroidal axis, it may be determined easily about any parallel axis. To prove this statement, consider the two parallel axes in Fig. 3.16, one being a centroidal axis through some other point C. The radial distances from the two axes to any element of mass dm are r_0 and r, and the separation of the axes is d. Substituting the law of cosines $r^2 = r_0^2 + d^2 + 2r_0 d \cos \theta$ into the definition for the moment of inertia about the noncentroidal axis through C gives

$$I = r^2 dm = \int (r_0^2 + d^2 + 2r_0 d \cos \theta) \, dm$$

$$= \int r_0^2 \, dm + d^2 \int dm + 2d \int u \, dm$$

The first integral is the moment of inertia $\bar{I}$ about the mass-center axis, the second term is md^2, and the third integral equals zero, since the u coordinate of the mass center with respect to the axis through G is zero. Thus the parallel-axis theorem is

$$I = \bar{I} + md^2 \tag{14}$$

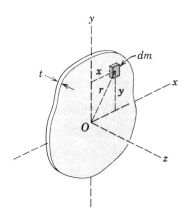

FIGURE 3.17

It must be remembered that the transfer cannot be made unless one axis passes through the center of mass and unless the axes are parallel. When the expressions for the radii of gyration are substituted in Eq. (14), there results

$$k^2 = \bar{k}^2 + d^2$$

which is the parallel-axis theorem for obtaining the radius of gyration k about an axis a distance d from a parallel controidal axis for which the radius of gyration is $\bar{k}$.

For plane-motion problems where rotation occurs about an axis normal to the plane of motion, a single subscript for I is sufficient to designate the inertia axis. Thus, if the plate of Fig. 3.17 has plane motion in the x-y plane, the moment of inertia of the plate about the z axis through O is designated I_o. For three-dimensional motion, however, where components of rotation may occur about more than one axis, a double subscript can be used to preserve notational symmetry with product-of-inertia terms. Thus the moments of inertia about the x, y, and z axes are labeled I_{xx}, I_{yy}, and I_{zz}, respectively, and from Fig. 3.17 it can be seen that they become

$$I_{xx} = \int r_x^2 \, dm = \int (y^2 + z^2) \, dm$$

$$I_{yy} = \int r_y^2 \, dm = \int (z^2 + x^2) \, dm$$

$$I_{zz} = \int r_z^2 \, dm = \int (x^2 + y^2) \, dm$$

These integrals are the same for angular momentum in three-dimensional rotation.

The similarity between the defining expressions for mass moments of inertia and area moments of inertia is easily observed. An exact relationship between the two moment-of-inertia expressions exist in the case of flat plates. Consider the flat plate of uniform thickness shown in Fig. 3.17. If the constant thickness is t and the density is ρ, the mass moment of inertia I_{zz} of the plate about the z axis normal to the plate is

$$I_{zz} = \int r^2 \, dm = \rho t \int r^2 \, dA = \rho t J_z$$

and equals the mass per unit area, ρt, times the polar moment of inertia J_z of the plate area about the z axis. If t is small compared with the dimensions of the plate in its plane, the mass moments of inertia I_{xx} and I_{yy} of the plate about the x and y axes are closely approximated by

$$I_{xx} = \int y^2 \, dm = \rho t \int y^2 \, dA = \rho t I_x$$
$$I_{yy} = \int x^2 \, dm = \rho t \int x^2 \, dA = \rho t I_y$$

(15)

Hence the mass moments of inertia equal the mass per unit area ρt times the corresponding area moments of inertia. The double subscripts for mass moments of inertia distinguish these quantities from area moments of inertia.

Inasmuch as $J_z = I_x + I_y$ for area moments of inertia, the equation

$$I_{zz} = I_{xx} + I_{yy}$$

(16)

which holds only for a thin flat plate. This restriction is observed from Eqs. (15), which do not hold true unless the thickness t or z coordinate of the element is negligible compared with the distance of the element from the corresponding x or y axis. Equation (16) is very useful when dealing with a differential mass element taken as a flat slice of differential thickness, say dz. In this case Eq. (16) holds exactly and becomes

$$dI_{zz} = dI_{xx} + dI_{yy}$$

3.1.13 Products of Mass Inertia

For problems in the rotation of three-dimensional rigid bodies the expression for angular momentum contains, in addition to the moment-of-mass inertia terms, product-of-mass inertia terms defined as

$$I_{xy} = I_{yx} = \int xy \; dm$$

$$I_{xz} = I_{zx} = \int xz \; dm$$

$$I_{yz} = I_{zy} = \int yz \; dm$$

Unlike moments of mass inertia, which are always positive quantities, products of mass inertia may be positive or negative. The calculation of products of mass inertia involves the same basic procedure as that used in calculating moments of mass inertia and in evaluating other volume integrals insofar as the choice of element and the limits of integration are concerned. The only special precaution is to be doubly watchful of the albebraic signs in the expressions. The units of products of mass inertia are the same as those of moments of mass inertia.

The calculations of moment of mass inertia is often simplified by using the transfer-of-axis theorem. A similar theorem exists for transferring products of mass inertia, and it can easily be proved as follows. Figure 3.18 shows the x-y view of a rigid body with parallel axes $x_0 y_0$ passing through the mass center G and located from the x-y axes by the distance d_x and d_y. The product of inertia about the x-y axes by definition is

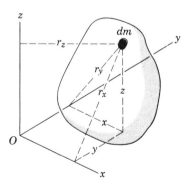

FIGURE 3.18

$$I_{xy} = \int xy \ dm = \int (x_0 + d_x)(y_0 + d_y) dm$$

$$I_{xy} = \int x_0 y_0 dm + d_x d_y \int dm + d_x \int y_0 \ dm + dy \int x_0 dm$$

$$I_{xy} = I_{x_0 y_0} + md_x dy$$

The last two integrals vanish since the first moments of mass about the mass center are necessarily zero. Similar relations exist for the remaining two product-of-mass inertia terms. If the zero subscripts are dropped and the bar is used to designate the mass-center quantity, the following equations are obtained:

$$I_{xy} = \overline{I}_{xy} + md_x d_y$$

$$I_{xz} = \overline{I}_{xz} + md_x d_z$$

$$I_{yz} = \overline{I}_{yz} + md_y d_z$$

These transfer-of-axis relations are valid only for transfer to or from parallel axes through the mass center.

3.1.14 External Surface Area

When the curve $y = f(x)$ is revolved about the X axis, a surface is generated as shown in Fig. 3.19. To find the area of this surface, the

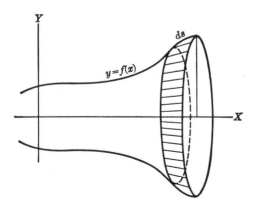

FIGURE 3.19

area generated by an element of arc ds is considered. This area is roughly that of a cylinder of radius y and $dS = 2\pi y\ ds$ can be written. Summing all such elements of surface area, the following equation is obtained:

$$S = 2\pi \int y\ ds$$

$$S = 2\pi \int_a^b y\ \sqrt{1 + y'^2}\ dx$$

Appropriate modifications of this formula will be necessary if the curve is revolved about some other line or polar coordinates are used and so on.

3.2 MASS PROPERTY CALCULATIONS (MPC) USING AN ICGS

The CADDS 4 Mass Property Calculations (MPC) package allows the user to calculate various mechanical properties of a surface or an object whose boundary is described by graphical entities (including lines, arcs, circles, B-splines, strings, and nodal lines). The user can calculate the following properties using MPC:

Volume or area
Mass
First moments
Center of mass
Moments and products of inertia with respect to the model space
    X, Y, and Z axes
Moment of inertia with respect to a given axis
Polar moment of inertia about a point
Radius of gyration
Principal axes
External surface area

The wire frame, surface edge, and solid models are the three types of models (three-dimensional objects) which can be created on most interactive computer graphics systems. A model can be defined by different geometric elements. If may be defined by points connected by lines, fy surfaces and edges or, finally, as a solid.

When a model is defined by points connected by lines, it is sometimes referred to as a *wire frame model*. Each line must be cut to simulate a section cut (sectioning produces four points), and only data concerning the intersection of linces can be obtained. If lines are desired between the resultant points, they must be constructed. For this model no surfaces exist; areas and volume cannot be obtained from existing information. Viewing requires a great deal of knowledge, and deductive reasoning and additional construction is time consuming.

The surface edge model is an object defined by surfaces and planes. For this model surfaces and edges can be intersected, distances can be extracted, and additional lines and edges can result from sectioning by a cutting plane. Surface areas and centers can be calculated and displayed. The volume of the object can be computed. Because the object is better defined, answers are easier to obtain and additional construction and editing are easier than with the wire frame model. The resulting section, however, is defined by four lines, not by a plane, and will not react to commands requiring a plane.

The solid model is an object defined as a solid. A solid may appear as a wire frame or as a shaded object, but it reacts to all commands with much greater intelligence. Mass property requests render the results expected, including weight if density input is allowed. A more intelligent system would also make adjustments for holes and cavities, but at present, this capability is found only in very refined systems that are programmed to understand names such as "tube" and "cylinder."

Throughout this chapter the surface edge model will be used for calculating the mass properties of an object because of computer processing restrictions.

3.2.1 Types of Objects for Which Mass Properties Commands (MPC) Are Defined

The mass properties are calculated for four types of objects, which are defined as follows.

1. A two-dimensional object is a planar region bounded by a series of curves (i.e., lines, arcs, conics, B-splines, strings, and nodal lines). The planar region can consist of both positive (solid) and negative (hole) components (see Fig. 3.20).

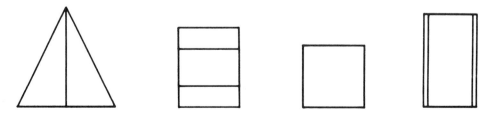

FIGURE 3.20 Two-dimensional object with holes.

FIGURE 3.21 Cone generated by a triangle.

Requirements: All boundary curves must lie in the same plane. The number of closed boundaries (solid and holes) is not limited, but the number of entities in each boundary must be less than 100.

2. A general surface of revolution is a three-dimensional object created by rotating one or more closed boundaries about an axis (any rotationally symmetrical object). A closed boundary consists of a series of curves (i.e., lines, arc, conics, B-splines, strings, and nodal lines). A boundary can enclose either a positive (solid) or a negative (hole) region. Examples of a general surface of revolution are: a cone generated by a triangle (see Fig. 3.21), a sphere generated by a closed half circle (Fig. 3.22), or a torus generated by a circle (Fig.

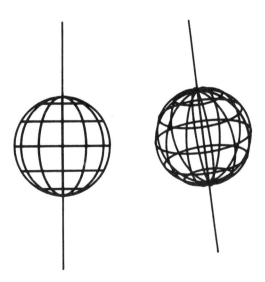

FIGURE 3.22 Sphere generated by a closed half circle.

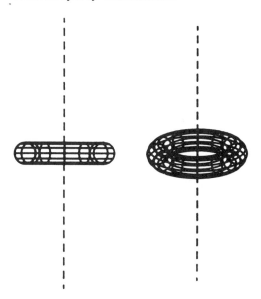

FIGURE 3.23 Torus generated by a circle.

3.23). Mass properties are calculated based on the closed boundary
from which the three-dimensional object is generated (such as the triangle, closed half circle, or circle in the examples above).

Requirements: All boundary curves must lie in the same plane.
The axis of revolution must lie in the same plane and cannot intersect
the curves, although the axis can touch the curves. The number of
closed boundaries (solid and holes) is not limited, but the number of
entities in each boundary must be less than 100.

3. A general tabulated cylinder is a three-dimensional object created
by projecting one or more closed boundaries about an axis. A closed
boundary consists of a series of curves (i.e., lines, arcs, conics, B-
splines, strings, and nodal lines). A boundary can enclose either a
positive (solid) or a negative (hole) region. Examples of a general
tabulated cylinder are: a cylinder generated by a circle (see Fig.
3.24), a pipe generated by two concentric circles (Fig. 3.25), and a
box generated by a square (Fig. 3.26). Mass properties are calculated
based on the closed boundary from which the three-dimensional object
is generated (such as the circle, two concentric circles, or square in
the examples above).

Requirements: All boundary curves must lie in the same plane.
The axis of projection must *not* lie in the same plane. The number of
closed boundaries (solid and holes) is not limited, but the number of
entities in each boundary must be less than 100.

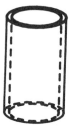

FIGURE 3.24 Cylinder   FIGURE 3.25 Pipe generated by
generated by a circle.   two concentric circles.

4. To calculate the mass properties of an object that is not a solid
of rotation or projection, the user can represent the object as a series
of cross sections. In this "cut and sum" method, mass properties are
calculated for each cross section and then summed to approximate the
mass properties of the object (see Fig. 3.27).

A cross section consists of one or more closed boundaries lying on
a plane. (A cross section can also consist of a single point.) A closed
boundary consists of a series of curves (i.e., lines, arcs, conics, B-
splines, strings, and nodal lines). A boundary can enclose either a
positive (solid) or a negative (hole) region.

A type IV object is typically treated as follows: (1) define the body
to be evaluated as a collection of CADDS 4 surfaces; and (2) using the
CUT PLANE command, generate the cross sections on a sequence of
layers. Alternatively, you can input the cross sections directly. The
mass properties of each cross section are then calculated and summed.

The calculation for each cross section is exact (within the precision
of the system) but the summation of the results of the cross sections
is approximate. Therefore, the accuracy of the results depends on
the number of cuts (cross sections). In general, a larger number of
cuts will yield better results.

FIGURE 3.26 Box generated by
a square.

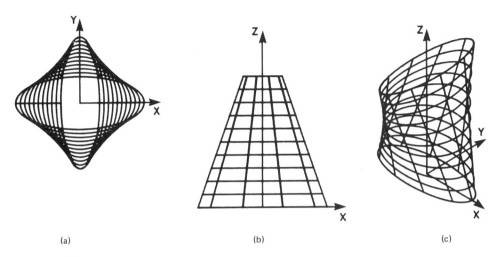

(a) (b) (c)

FIGURE 3.27 Arbitrary object defined by cross section: (a) top view; (b) front view; (c) isometric view.

Requirements: All boundary curves in each cross section must lie in the same plane. The distance between cross sections need not be fixed (i.e., variable thickness is allowed between cross sections). The user must exercise care to interpret the results obtained by the cut-and-sum method.

3.2.2 Procedures and an Overview of MPC Compounds

To begin working with the MPC package, the user must activate an existing part that has entities in it. Based on the shape of the object (one of the four types described in Section 3.2.1) for which the user wants to calculate mass properties, the user should determine the method that he or she will use for the calculations. Each method has a corresponding command modifier that is used. These modifiers/methods are shown in uppercase letters below.

PLANE: used to calculate the mass properties of a two-dimensional object
ROTATE: used to calculate the volume or mass properties of a general surface of revolution
PROJECT: used to calculate the volume or mass properties of a general tabulated cylinder
XSECTION: used to calculate the volume or mass properties of an arbitrary object defined by cross sections

The MPC commands used are determined by the type of object and method the user has chosen.

There are seven commands associated with MPC. Three of the commands calculate mass properties; four of the commands manipulate mass points which will not be covered. MPC commands operate at CADDS level in the Model mode (except the CALCULATE AREA command, which operates either in Model mode or Draw mode).

The three commands that calculate various mass properties are: CALCULATE AREA, CALCULATE VOLUME, and CALCULATE MASS PROPERTIES (MPROPERTIES).

> *CALCULATE AREA:* Computes the area inside two-dimensional boundaries defined on a plane.
>
> *CALCULATE VOLUME:* Computes the volume of three-dimensional objects defined in one of the following ways: (1) projection of boundaries along an axis, (2) rotation of boundaries about an axis, and (3) summation of a series of cross sections. It also computes mass and, optionally, external surface area.
>
> *CALCULATE MASS PROPERTIES (MPROPERTIES):* Computes the mass properties of four kinds of objects: a two-dimensional region defined by boundaries on a plane, and the three kinds of three-dimensional objects listed in the CALCULATE VOLUME description above. The calculated properties include volume or area, mass, center of mass, first moments, second moments, and optionally, external surface area.

These three commands will, if requested, store the calculated properties in a mass point entity which the user can access at any time.

The following format convention is used to describe each command.

Command: The complete command name is given under this heading.

Description: The description explains the general purpose and use of the command.

Valid modes: This section specifies the mode(s) in which the user can use the command (Model mode and/or Draw mode).

Syntax: The command syntax is given. The CADDS 4 command syntax involves two parts: (1) the command and, in most cases, (2) the location information and/or entity identification, which is referred to as *Getdata*. (Recall that anentity is any graphical element, such as a line, circle, arc, surface, dimension, text, and so on.)

The command consists of a verb-noun combination and modifiers that instruct the system to perform an action. Getdata is user-defined input that specifies what entities or locations will be affected. A colon (:) separates the command from Getdata. The command is to the left of the colon and Getdata is to the right. The user can read the command line as a sentence in which the colon means "as follows."

Symbolic notation is used to define the required and optional portions of the command. Symbols used in the command syntax are:

Uppercase A–Z Indicates command or Getdata information which the user should enter as shonw.

Lowercase Z–Z Indicates additional information which the user supplies (such as a filename or text).

Italicized A–Z, a–z Specifies the system prompt as it is displayed. Prompts generally ask for entity or location information.

< > Indicates required information which the user supplies (such as a filename). The angle brackets are not keyed in when entering the information.

[ ] Indicates optional modifiers. The modifiers are keyed in as listed in the command description. The square brackets are not keyed in when entering the modifer(s).

n Indicates a user-supplied integer, required with certain modifiers.

[CR] Indicates a return. A return is entered by pressing the RETURN key. This action instructs the system to process the command.

Modifiers (*dependent and independent*): This section lists and describes the modifiers that can be used in a command. Modifiers qualify the command action and are of two types: dependent and independent.

Dependent modifiers are grouped in a level hierarchy of associated modifiers: first level and second level. A first-level modifier may, in some cases, be required to use the command. In other cases, selecting a first-level modifier can be optional. Second-level modifiers further qualify the meaning or action of an associated first-level modifier. Therefore, a second-level modifier is dependent on and can only be used *after* the user has specified its associated first-level modifier. This association is outlined in a table in the Modifiers sections for each command description. (Note that not every command includes levels of modifiers; therefore, not every command description includes a table.)

A portion of the modifier table for the CALCULATE VOLUME command is shown here as an example of the level association.

First-level modifiers (the user must choose only one)	Second-level modifiers (optional)
PROJECT	LOWBND HIBND
ROTATE	LOWANG HIANG

Notice that LOWBND and HIBND are the second-level modifiers associated with PROJECT. If PROJECT is used, the user can choose only LOWBND and/or HIBND as second-level modifiers. Similarly, ROTATE is used, the user may choose LOWANG and/or HIANG.

A description of each modifier follows each table. In the modifier descriptions, the levels are indicated at the left margin as *1* or *2*.

Independent modifiers are not level dependent. In most cases, their use is optional. Some commands have independent modifiers only and do not require the user to specify a modifier at all. Other commands require the user to specify at least one independent modifier. Each independent modifier is explained in the Modifiers section. If a command has dependent and independent modifiers which the user wants to use, the dependent modifiers are entered first (in their proper order) and then the independent modifiers.

Procedure: The procedure explains the proper sequence of input of the basic command (verb and noun, without modifiers in some cases), the system's response (or prompts), and the Getdata input. Note that when instructed by the procedure to digitize a location or entity, the user can either digitize on the tablet or enter explicit coordinates, in most cases.

Examples: Examples graphically illustrate and clarify the command and its variations using modifiers.

Notes: Notes provide special information about using the command.

3.3   MASS PROPERTIES CALCULATION COMMANDS

Three CADDS 4 commands are used to calculate mass properties: CAL-CULATE AREA, CALCULATE VOLUME, and CALCULATE MPROPER-TIES. These commands and their associated modifiers are described in this section.

3.3.1 Calculate Area Command

Command

CALCULATE AREA

Description

The CALCULATE AREA command computes the area inside one or more closed boundaries for a two-dimensional object, optionally subtracting the area of one or more holes if any exist. Solid and hole boundaries may consist of lines, arcs, conics, B-splines, strings, and nodal lines. If the COORDINATE modifier is specified, solid and hole boundaries are defined by digitizing locations which the system connects by lines to form a polygon which stores the calculated area is created, if requested.

Valid Modes

Model mode or Draw mode. In the Model mode, the user can identify only model entities. In the Draw mode, the user can identify only drawing entities.

Syntax

#n# CALCULATE AREA (modifiers):

Modifiers

Only independent modifiers are included in this command. These modifiers can be used optionally.

Independent Modifiers

HOLE: If specified, the system will prompt the user for identification of one or more hold boundaries (the negative area) after
the user identifies the solid boundaries (the positive area). The
hole area is subtracted form the total positive area.

COORDINATE: Indicates that boundaries will be defined by digitizing locations that act as the corners (vertices) of a polygon.
The user would use this modifier to calculate the area of a region that is not bounded by existing entities. The system closes
the polygon, connecting the first and last points automatically.
The user must digitize at least three points per boundary.

In the discussion that follows, we are going to assume that the
reader is now a user, and address you directly.

Procedure

1. Key in CALCULATE AREA and any of the modifiers listed, if de-
    sired. Key in a colon. The system prompts *PLEASE ENTER
    SOLID BOUNDARY #1:*. Digitize to identify entities in order (one
    after another) around a boundary; then enter a return.
2. The system prompts for another set of boundary entities, *PLEASE
    ENTER SOLID BOUNDARY #2:*. If the area being calculated is
    defined by more than one boundary, repeat the entity identifica-
    tion (digitizing) procedure.

At this point, if you did not specify a modifier (or used only the
MPOINT modifier), continue with step 3a. If you used the HOLE modifier, continue with step 3b. If you used the COORDINATE modifier,
continue with step 3c.

3a. If no modifier was selected, the system continues to prompt for
    boundaries until you enter a return without identifying entities.
    This return executes the command (unless you use the HOLE
    modifier, as described in step 3b). The system then displays
    the calculated value(s) for the area(s).
3b. If you use the HOLE modifier, the system prompts you with
    PLEASE ENTER HOLE #1: after the solid boundaries are defined.
    Digitize to define the boundary entities for the first hole; then

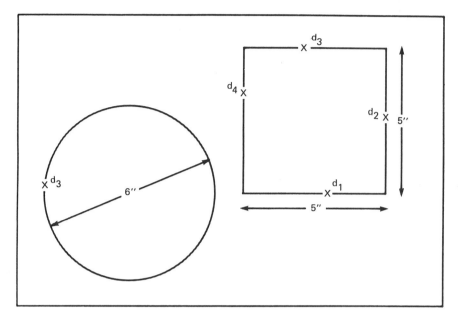

FIGURE 3.28 CALCULATE AREA—no modifiers are used.

enter a return. The system prompts you again to identify another hole boundary. Follow the same procedure if this is desired; otherwise, key in a return to execute the command. The
system calculates and displays the solid area minus the area of
the hole(s).

3c. If you want to calculate the area of a region that is not bounded
by existing entities, use the COORDINATE modifier. The system
allows you to directly digitize locations to define a solid or hole
boundary. The locations form a polygon, with each digitize defining a point (corner) of the polygon. Lines are graphically
displayed between the first and last location defined, although a
connecting last line is not displayed on the screen.

Example 1

Calculate the area of a square (boundary 1) and a circle (boundary
2) (see Fig. 3.28).

#n# CALCULATE AREA:

PLEASE ENTER SOLID BOUNDARY #1: $d_1d_2d_3d_4$ [CR]
PLEASE ENTER SOLID BOUNDARY # 2: d_5 [CR]

PLEASE ENTER SOLID BOUNDARY # 3: [CR]
AREA   = 0.5327433E+02

Mathematical Solution

Area of circle = πr^2 = $(3)^2$ 28.27433388 in.^{2}

Area of square = a(b) = 5(5)   + 25.0   in.^{2}

————————————————————

   Total of area of circle and square =   53.27433388 in.^{2}

Example 2

Calculate the area of a square with two holes; create a mass point (centroid) and store the calculated values (see Fig. 3.29).

#n# CALCULATE AREA HOLE MPOINT:

PLEASE ENTER SOLID BOUNDARY # 1: $d_1 d_2 d_3 d_4$ [CR]

PLEASE ENTER SOLID BOUNDARY # 2: [CR]

PLEASE ENTER HOLE # 1: d_5 [CR]

PLEASE ENTER HOLE # 2: d_6 [CR]

PLEASE ENTER HOLE # 3: [CR]

AREA   = 0.2342920E+02

CENTROID

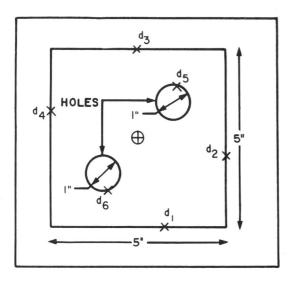

FIGURE 3.29 CALCULATE AREA using the hole modifier.

$X = 0.2500000E+01$ $Y = 0.2500000E+01$ $Z = 0.0000000E+00$

Mathematical Solution

Area of square $= a(b) = 5(5)$ $=$ 25.000000000 in.^{2}
Area of holes $= 2(\pi r^2) = 2\pi(0.5)^2 =$ -1.570796327 in.^{2}

Total area $=$ 23.429203670 in.^{2}

$$A_t \bar{x} = \sum_1^3 A_n x_n$$

$23.42920367\bar{x} = 25(2.5) - 0.7853981635(1.5) - 0.7853981635(3.5)$

$$\bar{y} = \frac{62.5 - 1.178097245 - 2.748893572}{23.42920367}$$

$= 2.5$ in.

$$A_t \bar{y} = \sum_1^3 A_n y_n$$

$23.42920367\bar{y} = 25(2.5) - 0.7853981635(1.5) - 0.7853981635(3.5)$

$$\bar{y} = \frac{62.5 - 1.178097245 - 2.748893572}{23.42920367}$$

$= 2.5$ in.

3.3.2 Calculate Volume Command

Command

CALCULATE VOLUME

Description

The CALCULATE VOLUME command computes the volume (and optionally the surface area) of an object defined by one of the following methods:

The projection of planar boundaries along an axis
The rotation of planar boundaries about an axis
A series of planar cross sections

Boundaries may consist of lines, arcs, conics, B-splines, strings, and nodal lines. The boundaries may represent solid regions or holes.

Valid Modes

Model mode only.

Syntax

#n# CALCULATE VOLUME <PROJECT or ROTATE or XSECTION>
(additional modifiers):

Modifiers

This command includes dependent and independent modifiers. The system requires that you choose one, and only one, first-level modifier. You then can select an associated second-level modifier (s). Independent modifiers are optional.

CALCULATE VOLUME Modifier Table

This table presents all modifiers available with the CALCULATE VOLUME command. Default values for modifiers are given in parentheses. A description of each modifier follows the table.

Dependent Modifiers

First-Level Modifiers (you must choose only *one*)	Second-Level Modifiers (optional)
PROJECT	LOWBND n (0.0) HIBND n (1.0)
ROTATE	LOWANG n (0.0) HIANG n (360.0)
XSECTION	NSTEP n (2) LAY n (active construction layer) ILAY n (1)

Independent Modifiers

Choose in addition to those listed above:

    HOLE
    SAREA
    DENSITY n (1.)

Dependent Modifiers

1 PROJECT Indicates that the object for which volume will be calculated is defined by the projection of boundaries along an axis.
2 LOWBND n Specifies the distance from the plane of the boundaries to the bottom (lower limit) of the projected object; measured along the axis from the plane of the boundaries (Fig. 3.30).
2 HIBND n Specifies the distance from the plane of the boundaries to the top (upper limit) of the projected object; measured along the axis from the plane of the boundaries (see Fig. 3.30). Default is 1.0.

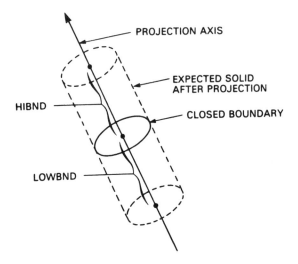

FIGURE 3.30 Illustration showing how HIBND and LOWBND are measured.

1 ROTATE Indicates that the object for which volume will be calculated is defined by the rotation of boundaries about an axis.

2 LOWANG n Specifies the minimum extent of the rotation; measured in degrees from the plane of the boundaries according to to the right-hand rule.
Default is 0.0.

2 HIANG n Specifies the maximum extent of the rotation; measured in degrees from the plane of the boundaries according to the right-hand rule.
Default is 360.0.

1 XSECTION Indicates that the object for which volume will be calculated is defined by a series of cross sections. If either the LAY or ILAY second-level modifiers are selected, it is assumed that each section is on a different layer, and the layers are echoed sequentially. Otherwise, the visible layers remain unchanged.

2 Specifies the number of cross sections defining the object.
Default is 2.

2 LAY n Specifies the layer number on which you will digitize the first cross section.
Default is the active construction layer.

2 ILAY n Specifies the increment in the layer number of successive cross sections.
Default is 1.

Independent Modifiers

HOLE If this modifier is specified, the system prompts you for
identification of hole boundaries after you identify the solid boundaries.

SAREA Calculates the outermost surface area of the object (i.e.,
for ROTATE and PROJECT, the area generated by the first boundary digitized; for XSECTION, the area of a surface interpolated
between the first boundaries of each section).

DENSITY n Specifies the density of the object.
Default is 1.0.

Procedure

1. Key in CALCULATE VOLUME followed by *one* of the first-level
   modifiers (PROJECT, ROTATE, or XSECTION).
2. Key in associated second-level modifiers and/or independent modifiers, if desired. Enter a colon. At this point, if you used the
   PROJECT or ROTATE modifier, continue with steps 3a, 4a, and 5a.
   If you used the XSECTION modifier, continue with steps 3b, 4b,
   and 5b.
3a. If you use the PROJECT or ROTATE modifiers, the system prompts
   *PLEASE DIGITIZE TWO POINTS TO DEFINE PROJECTION (or RO-
   TATION) AXIS.* The *MODEL loc* prompt is then displayed on the
   next line. Digitize two locations to identify an axis for projection or rotation. Enter a return.
4a. The system then prompts *PLEASE ENTER SOLID BOUNDARY #1:.*
   Digitize to identify the entities defining the planar boundaries to
   be projected or rotated about the axis. Always digitize these entities in order (one after another) around a boundary. Enter a
   return.
5a. The system then prompts *PLEASE ENTER SOLID BOUNDARY #2:,*
   requesting that you identify another set of boundary entities.
   If the object is defined by more than one boundary, repeat the
   digitizing procedure (as in step 4a); otherwise, enter a return
   to execute the command. The system displays the calculated
   value for the volume.
3b. If you use the XSECTION modifier, the system prompts *CROSS-
   SECTION #1* and, on the next line, *PLEASE ENTER SOLID BOUN-
   DARY #1:.* Digitize to identify the entities defining the boundary of the first planar cross section. Enter a return.
4b. The system then prompts *PLEASE ENTER SOLID BOUNDARY #2:,*
   requesting that you identify another set of boundary entities
   within cross section #1. If the cross section is defined by more
   than one boundary, repeat the digitizing procedure. Continue
   this process until cross section #1 is completed. Enter a return.

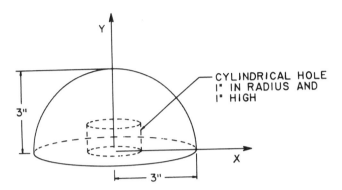

FIGURE 3.31 Geometry of the object: a hemisphere with a cylindrical hole in the base.

5b. The system then displays *CROSS-SECTION #2* and prompts you to identify another set of boundary entities. Follow the same procedure as for cross section #1 in step 4b. Repeat this process as many times as there are cross sections which define the object. The system displays the calculated value for the volume.

Example 3

Calculate the volume of a hemisphere with a cylindrical hole in the base. The geometry of the object is shown in Fig. 3.31. Figure 3.32 presents the example as it would appear on the screen.

#n# CALCULATE VOLUME ROTATE SAREA MPOINT:

PLEASE DIGITIZE TWO POINTS TO DEFINE ROTATION AXIS

MODEL loc d_1d_2 *[CR]*

PLEASE ENTER SOLID BOUNDARY # 1: $d_3d_4d_5d_6d_7$ *[CR]*
PLEASE ENTER SOLID BOUNDARY # 2: [CR]

VOLUME	$= 0.5340404E+02$
MASS	$= 0.5340707E+02$
SURFACE AREA	$= 0.9110618E+02$

CENTROID:
    $X = -0.2780463E-06$ $Y = 0.1161765E+01$ $Z = 0.0000000E+00$
FIRST MOMENTS:
$FX = -0.1484964E-04$ $FY = 0.6204645E+02$ $FZ = 0.0000000E+00$
MOMENTS OF INERTIA:
$IXX = 0.2017426E+03$ $IYY = 0.2020044E+03$ $IZZ = 0.2017426E+03$

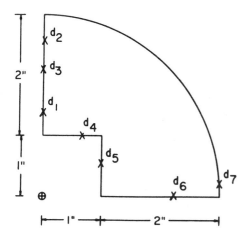

FIGURE 3.32 CALCULATE MPROPERTIES using the ROTATE modifier.

PRODUCTS OF INERTIA:
$IXY = -0.1529058E-04$ $IXZ = -0.146201eE-10$ $IYZ = 0.0000000E+00$

Mathematical Solution

Volume
Volume of hemisphere = V_1 = positive volume

$$dV_1 = (\ - y^2)dy$$

$$V_1 = \pi \int_0^3 (9 - y^2)dy = \pi\left[9y - \frac{y^3}{3}\right]_0^3 = \pi(27 - 9) = \pi(18)$$

$$= 56.54866776 \text{ in.}^3$$

Volume of cylinder = V_2 = negative volume

$$dV_2 = -\pi r^2 \, dy$$

$$V_2 = -\pi r^2 \int_0^1 dy = \pi r^2 y \,\Big|_0^1 = \pi(1)^2(1)$$

$$= -3.141592654 \text{ in.}^3$$

$$V_1 + (-V_2) = 56.54866776 - 3.141592654 = 53.4070754 \text{ in.}^3$$

Default $\sigma = 1$ lb/in.^{3}. Therefore, mass = 53.4070754 lb.

Surface area

Hemisphere

$$S_1 = a^2 \int_0^\pi d\theta \int_0^\pi \sin\phi \, d\phi$$

$$= a^2 [\theta]_0^\pi [-\cos\phi]_0^\pi = a^2\pi^2 = 2\pi a^2$$

$$= 2\pi(3)^2 = 56.54866776 \text{ in.}^2$$

Base

$$S_2 = \pi r_1^2 - \pi r_2^2 = \pi[3^2 - 1^2] = \pi 8 = 25.13274123 \text{ in.}^2$$

Hole

$$S_3 = 2\pi r + \pi r_2^2 = 2\pi(1) + \pi(1)^2$$

$$= 6.283185307 + 3.141592654 = 0.424777961 \text{ in.}^2$$

Total Surface Area

$$S_T = S_1 + S_2 + S_3 = 56.54866776 + 25.13274123 + 9.424777961$$

$$= 91.10618695 \text{ in.}^2$$

Centroid

By symmetry, $\bar{x} = 0$ and $\bar{z} = 0$.

Volume of hemisphere = 56.54866776 in.^{3}

Volume of cylinder   = -3.141592654 in.^{3}
_____

Total of volume   = 53.4070754 in.^{3}

Centroid of hemisphere

$$dV_1 = \pi(r^2 - y^2)dy$$

$$V_1\bar{y} = \int y_c dV$$

$$56.54866776\overline{y} = \int_0^3 y\pi(9 - y^2)dy = \pi\left[\frac{y^2}{2}(9) - \frac{y^4}{4}\right]_0^3$$

$$= \pi\left[\frac{81}{2} - \frac{81}{4}\right] = \pi\left[\frac{81}{4}\right]$$

$$\overline{y} = 1.125 \text{ in.}$$

Centroid of cylinder

By symmetry, $\overline{y} = 0.5$ in.

$$V_1\overline{y} = \sum_1^2 V_n y_n$$

$$53.4040754\overline{y} = 56.54866776(1.125) + (-3.141592654)0.5$$
$$= 63.61725123 - 1.570796327$$

$$\overline{y} = \frac{62.0464549}{53.4070754} = 1.161764699 \text{ in.}$$

Moment of Inertia

Hemisphere

$$dI_{yy_1} = \frac{1}{2}(dm)y^2 = \frac{1}{2}(\pi\rho y^2 dx)y^2$$

$$= \frac{\pi\rho}{2}(r^2 - x^2)dx$$

$$I_{yy_1} = \frac{\pi\rho}{2}\int_0^r (r^2 - x^2)^2 dx = \frac{\pi\rho}{2}\left[r^4 x - \frac{r^4 x - 2r^2 x^3}{3} + \frac{x^5}{5}\right]_0^r$$

$$= \frac{8}{30\pi\rho r^5} = \frac{8}{30\pi\rho r^5}$$

If $\rho = 1$, then

$$I_{yy_1} = 4/15\pi r^5$$

and if $r = 3$, then

$$I_{yy_1} = \frac{4}{15\pi(3)^5} = 203.575204 \text{ in.}^4$$

Cylinder

$$I_{yy_2} = \int r_0^2 \, dm = \rho t \int_0^{2\pi} \int_0^r r_0^3 \, dr_0 d\theta = \frac{\rho t \pi r^4}{2}$$

$$= \frac{1}{2} mr^2$$

If $m = v\rho$ and $\rho = 1$, then $m = V$.

$$I_{yy_2} = \frac{1}{2} Vr^2 = \frac{1}{2}(3.141592654)(1)^2 = 1.570796327 \text{ in.}^4$$

where

$$V = \pi r^2 h = \pi(1)^2(1) = 3.141592654 \text{ in.}^3$$

$$I_{yy \, T} = I_{yy_1} - I_{yy_2} = 203.575204 - 1.570796327$$

$$= 202.0044077 \text{ in.}^4$$

Moment of Inertia

Hemisphere

$$I_{xx_1} = 203.575204 \text{ in.}^4$$

Cylinder

$$I_{xx_2} = I_{xx}|_{c.g.} + m\left(\frac{h}{2}\right) = \left[\frac{1}{4} mr^2 + \frac{1}{12} mh^2\right] + m\frac{h^2}{2}$$

$$= \frac{1}{4} mr^2 + \frac{1}{3} ml^2 =$$

$$= \frac{1}{4}(3.141592654)(1)^2$$

$$+ \frac{1}{3}(3.141592654)(1)^2$$

Again, $m = V$.

$$I_{xx_2} = 0.7853981634 + 1.047197551$$

$$= 1.832595715 \text{ in.}^4$$

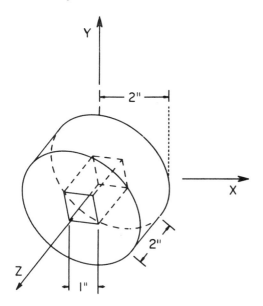

FIGURE 3.33 Geometry of the object to be measured using the CAL-CULATE MPROPERTIES command: a cylinder with a square hole along its axis.

$$I_{xx}\big|_T = I_{xx_1} - I_{xx_2} = 203.575204 - 1.832595715$$

$$I_{zz}\big|_T = I_{xx}\big|_T = 201.7426083 \text{ in.}^4$$

Example 4

Calculate the volume of a cylinder with a square hole along its axis. The geometry of the object is shown in Fig. 3.33. Figure 3.34 presents the example as it would appear on the screen.

#n# CALCULATE VOLUME PROJECT HIBND 2 HOLD SAREA MPOINT:

PLEASE DIGITIZE TWO POINTS TO DEFINE PROJECTION AXIS
MODEL loc X0Y0Z1 [CR]

PLEASE ENTER SOLID BOUNDARY # 1: d_1 *[CR]*
PLEASE ENTER SOLID BOUNDARY # 2: [CR]
PLEASE ENTER HOLE # 1: d_2 *[CR]*
PLEASE ENTER HOLE # 2: [CR]

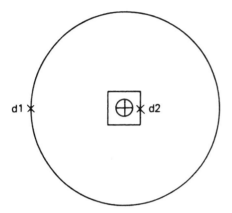

FIGURE 3.34 CALCULATE MPROPERTIES using the PROJECT modifier.

```
VOLUME           = 0.2313274E+02
MASS             = 0.2313274E+02
SURFACE AREA     = 0.4826548E+02
CENTROID:
X = 0.0000000E+00      Y = −0.3145307E-12      Z = 0.1000000E+01
FX = 0.0000000E+00     FY = −0.7275958E-11     FZ = 0.2313274E+02
MOMENTS OF INERTIA:
IXX = 0.5580974E+02    IYY = 0.5580977E+02     IZZ = 0.4993220E+02
PRODUCTS OF INERTIA:
IXY = −0.7275958E-11   IXZ = −0.0000000E+00    IYZ = −0.7275958E-11
```

Mathematical Solution

Volume

V_1 = volume of cylinder = $\pi r^2 h$ = $(2)^2(2)$ = 25.13274123 in.^{3}

V_2 = volume of square hole = (a)(b)(c) = (1)(1)(2) = 2 in.^{3}

V_T = $V_1 - V_2$ = 25.13274123 − 2.0 = 23.13274123 in.^{3}

and ρ = 1; thus

Mass = $V\rho$ = 23.13274123(1) = 23.13274123 in.^{3}

Surface Area

$$\text{Surface area} = 2[\pi r^2 - a(b)] + 2\pi r(h)$$

$$= 2[(\pi 2^2) - 1(1)] + 2\pi(2)2$$

$$= 23.13274123 + 25.13274123$$

$$= 48.26548246 \text{ in.}^2$$

Centroid

$$V_T \bar{z} = \sum_1^2 V_n z_n$$

$$23.13274123\bar{z} = 25.13274123(1) - 2(1)$$

$$\bar{z} = \frac{23.13274123}{23.13274123} = 1$$

$$23.13274123\bar{x} = 25.13274123(0) - 2(0) = 0$$

$$\bar{x} = 0$$

$$23.13274123\bar{y} = 25.13274123(0) - 2(0) = 0$$

$$\bar{y} = 0$$

Example 5

Calculate the volume of a section of a pyramid with a cylindrical hole. Figure 3.35 shows the geometry of the object and the cuts that generated the cross sections used in the mass properties calculations. Figure 3.36 presents the cross sections as they would appear on the screen. The system displays one cross section at a time because in this example each cross section has been placed on a different layer.

#n# CALCULATE VOLUME XSECTION LAY 4 ILAY 2 NSTEP 6 HOLE MPOINT SAREA :

CROSS-SECTION # 1

PLEASE ENTER SOLID BOUNDARY # 1: $d_1 d_2 d_3 d_4$ *[CR]*
PLEASE ENTER SOLID BOUNDARY # 2: *[CR]*
PLEASE ENTER HOLE # 1: d_5 *[CR]*
PLEASE ENTER HOLE # 2: *[CR]*

CROSS-SECTION # 2

PLEASE ENTER SOLID BOUNDARY # 1: $d_6 d_7 d_8 d_9$ *[CR]*
PLEASE ENTER SOLID BOUNDARY # 2: *[CR]*
PLEASE ENTER HOLE # 1: d_{10} *[CR]*
PLEASE ENTER HOLE # 2: *[CR]*

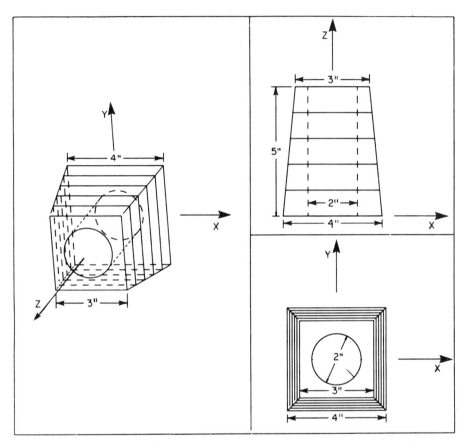

FIGURE 3.35 Geometry of the object to be measured using the CAL-CULATE VOLUME command: a section of a pyramid with a hole in the center.

CROSS-SECTION # 3

PLEASE ENTER SOLID BOUNDARY # 1: $d_{11}d_{12}d_{13}d_{14}$ *[CR]*
PLEASE ENTER SOLID BOUNDARY # 2: *[CR]*
PLEASE ENTER HOLE # 1: d_{15} *[CR]*
PLEASE ENTER HOLE # 2: *[CR]*

CROSS-SECTION # 4

PLEASE ENTER SOLID BOUNDARY # 1: $d_{16}d_{17}d_{18}d_{19}$ *[CR]*
PLEASE ENTER SOLID BOUNDARY # 2: *[CR]*
PLEASE ENTER HOLE # 1: d_{20} *[CR]*
PLEASE ENTER HOLE # 2: *[CR]*

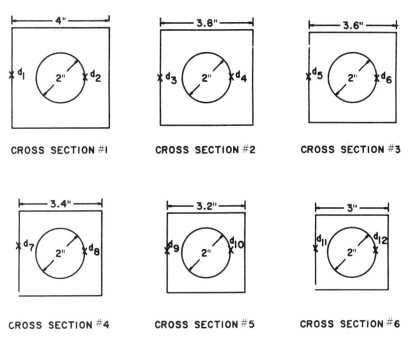

FIGURE 3.36 CALCULATE VOLUME using the XSECTION modifier.

CROSS-SECTION # 5

PLEASE ENTER SOLID BOUNDARY # 1: $d_{21}d_{22}d_{23}d_{24}$ [CR]
PLEASE ENTER SOLID BOUNDARY # 2: [CR]
PLEASE ENTER HOLE # 1: d_{25} [CR]
PLEASE ENTER HOLE # 2: [CR]

CROSS-SECTION # 6

PLEASE ENTER SOLID BOUNDARY # 1: $d_{26}d_{27}d_{28}d_{29}$ [CR]
PLEASE ENTER SOLID BOUNDARY # 2: [CR]
PLEASE ENTER HOLE # 1: d_{30} [CR]
PLEASE ENTER HOLE # 2: [CR]

VOLUME	= 0.4594673E+02
MASS	= 0.4594673E+02
SURFACE AREA	= 0.8906595E+02
CENTROID:	

X = 0.1006903E-06   Y = 0.7002763E-07   Z = 0.2182522E+01
FIRST MOMENTS:
FX = 0.4626389E-05   FY = 0.3217541E-05   FZ = 0.1002797E+03

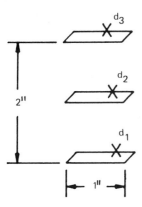

FIGURE 3.37 Three cross sections of a rectangular box.

MOMENTS OF INERTIA:
IXX = 0.3751889E+03   IVY = 0.3751889E+03   IZZ = 0.1227238E+03
PRODUCTS OF INERTIA:
IXY = 0.9830589E-05   IXZ = 0.1314150E-04   IYZ = 0.9586305E-05

Example 6

Calculate the volume of a 1 by 1 by 2 box without the use of layer echoing. The geometry of the object is shown in Fig. 3.37.

#n# CALCULATE VOLUME XSECTION NSTEP 3 SAREA MPOINT :

CROSS-SECTION # 1

PLEASE ENTER SOLID BOUNDARY # 1: d_1 [CR]
PLEASE ENTER SOLID BOUNDARY # 2: [CR]

CROSS-SECTION # 2

PLEASE ENTER SOLID BOUNDARY # 1: d_2 [CR]
PLEASE ENTER SOLID BOUNDARY # 2: [CR]

CROSS-SECTION # 3

PLEASE ENTER SOLID BOUNDARY # 1: d_3 [CR]
PLEASE ENTER SOLID BOUNDARY # 2: [CR]

VOLUME   = 0.2000000E+01
MASS   = 0.2000000E+01
SURFACE AREA   = 0.1000001E+02

CENTROID:
X = 0.5000000E+00   Y = 0.5000000E+00   Z = 0.1000000E+01
FIRST MOMENTS:
FX = 0.1000000E+01   FY = 0.1000000E+01   FZ = 0.2000000E+01
MOMENTS OF INERTIA:
IXX = 0.3333333E+01   IYY = 0.3333333E+01   IZZ = 0.1333333E+01
PRODUCTS OF INERTIA:
IXY = 0.5000000E+00   IXZ = 0.1000000E+01   IYZ = 0.1000000E+01

Notes

1. The axis for rotation *must* lie in the plane of the boundaries. The axis for projection *must not* lie in the plane of the boundaries. It is up to you to check that all of the boundaries lie on a single plane and that no boundary intersects itself.
2. If you specify a boundary incorrectly, use the US-DEL key or RUB OUT key to delete the last boundary defined (before another boundary is defined). You can use US-DEL or RUB OUT while defining a boundary to delete the last entity entered. If more deletions are entered than entities, the *entire* previous boundary defined is deleted.
3. When identifying the entities in a boundary, *always* digitize the entities in order (one after another) around a boundary.
4. You should select an empty layer (using the SELECT LAYER command) when using CALCULATE VOLUME XSECTION with layer echoing; otherwise, undesired graphics may remain on the screen.

3.3.3 Calculate Mass Properties Command

Command

CALCULATE MPROPERTIES

Description

The CALCULATE MPROPERTIES command computes the mass properties (volume or area, mass, centroid, first moments, moments of inertia, products of inertia, and optionally, surface area) of an object defined by one of the following methods:

One or more boundaries in a plane
The projection of planar boundaries along an axis
The rotation of planar boundaries about an axis
A series of planar cross sections

Boundaries may consist of lines, arcs, conics, B-splines, strings, and nodal lines. They may represent solid regions or holes. If requested, a mass point is created at the centroid of the object to store the calculated results. Values are given relative to model space coordinates.

Valid Modes

Model mode only.

Syntax

#n# CALCULATE MPROPERTIES <PLANE OR PROJECT or ROTATE or XSECTION> [additional modifiers]:

Modifiers

This command includes dependent and independent modifiers. The system requires that you choose one, and only one, first-level modifier. You then can select an associated second-level modifier(s). Independent modifiers are optional.

CALCULATE MPROPERTIES Modifier Table

This table presents all modifiers available with the CALCULATE MPROPERTIES command. Default values for modifiers are given in parentheses. A description of each modifier follows the table.

Dependent Modifiers

First-Level Modifiers (you must choose only *one*)	Second-Level Modifiers (optional)
PLANE	
PROJECT	LOWBND n (0.0) HIBND n (1.0)
ROTATE	LOWANG n (0.0) HIANG n (360.0)
XSECTION	NSTEP n (2) LAY n (active construction layer number) ILAY n (1)

Independent Modifiers

Choose in addition to those listed above:

HOLE
SAREA
DENSITY n (1.0)

Dependent Modifiers

1 PLANE   Indicates that the object for which mass properties will
be calculated is defined by boundaries in a plane. To maintain
the conventional meanings of the moments and products of inertia, the object should lie on the model space X-Y plane.

1 PROJECT Indicates that the object for which mass properties will
be calculated is defined by the projection of boundaries along an
axis.

2 LOWBND n Specifies the distance from the plane of the boundaries to the bottom (lower limit) of the projected body; measured
along the axis from the plane of the boundaries.
Default is 0.0.

2 HIBND n Specifies the distance from the plane of the boundaries
to the top (upper limit) of the projected body; measured along
the axis from the plane of the boundaries.
Default is 1.0.

1 ROTATE Indicates that the object for which mass properties will
be calculated is defined by the rotation of boundaries about an
axis.

2 LOWANG n Specifies the minimum extent of the rotation; measured in degrees from the plane of the boundaries according to
to the right-hand rule.
Default is 0.0.

2 HIANG n Specifies the maximum extent of the rotation; measured
in degrees from the plane of the boundaries according to the
right-hand rule.
Default is 360.0.

1 XSECTION Indicates that the object of which mass properties will
be calculated is defined by a series of cross sections. If the LAY
and/or ILAY second-level modifiers are selected, it is assumed
that each section is on a different layer, and the layers are echoed sequentially; otherwise, the visible layers remain unchanged.

2 NSTEP n Specifies the number of cross sections defining the object.
Default is 2.

2 LAY n Specifies the layer number on which you will digitize the
first cross section.
Default is your active construction layer.

2 ILAY n Specifies the increment in the layer number of successive
cross sections.
Default is 1.

Independent Modifiers

HOLE Specifies that the system will prompt for identification of
hole boundaries after you identify the solid boundaries.

SAREA Calculates the outermost surface area of the object (i.e., for ROTATE and PROJECT, the area generated by the first boundary digitized; and for XSECTION, the area of a surface interpolated between the first boundaries of each section).
DENSITY n Specifies the density of the object. Default is 1.0.
MPOINT Creates a mass point entity in which to store the calculated results.

Procedure

1. Key in CALCULATE MPROPERTIES followed by one of the first-level modifiers (PLANE, PROJECT, ROTATE, or XSECTION).
2. Key in associated second-level modifiers and/or independent modifiers, if desired. Enter a colon. At this point, ifyou used the PROJECT or ROTATE modifier, continue with steps 3a, 4a, and 5a. If you used the XSECTION modifier, continue with steps 3b, 4b, and 5b. If you used the PLANE modifier, continue with steps 3c and 4c.
3a. If you use the PROJECT or ROTATE modifier, the system prompts *PLEASE DIGITIZE TWO POINTS TO DEFINE PROJECTION (or ROTATION) AXIS.* The *MODEL loc* prompt then is displayed on the next line. Digitize two locations to identify an axis for projection or rotation. Enter a return. (See Examples 1 and 2.)
4a. The system then prompts *PLEASE ENTER SOLID BOUNDARY #1:.* Digitize to identify the entities defining the planar boundaries to be projected or rotated about the axis. Always digitize these entitites in order (one after another) around a boundary. Enter a return.
5a. The system then prompts *PLEASE ENTER SOLID BOUNDARY #2:,* requesting that you identify another set of boundary entities. If the object is defined by more than one boundary, repeat the digitizing procedure (as in step 4a); otherwise, enter a return to execute the command. The system displays the calculated value for the volume.
3b. If you use the XSECTION modifier, the system prompts *CROSS-SECTION #1* and, on the next line, *PLEASE ENTER SOLID BOUN-DARY #1:.* Digitize to identify the entities defining the boundary of the first planar cross section. Enter a return.
4b. The system then prompts *PLEASE ENTER SOLID BOUNDARY #2:,* requesting that you identify another set of boundary entities within cross section 1. If the cross section is defined by more than one boundary, repeat the digitizing procedure. Continue this process until cross section 1 is completed. Enter a return.
5b. The system then displays *CROSS-SECTION #2* and prompts you to identify another set of boundary entities. Follow the same procedure as for cross section 1 in step 3b. Repeat this pro-

cess as many times as there are cross sections which define the object. The system displays the calculated value for the volume.

3c. If you use the PLANE modifier, the system prompts *PLEASE ENTER SOLID BOUNDARY #1:.* Digitize to identify the boundary entities, then enter a return.

4c. The system prompts for another boundary identification. To calculate the mass properties of an object defined by more than one boundary, repeat the digitizing procedure (as in step 3c); otherwise, enter a return. The system displays the calculated mass properties values.

Example 7

Calculate the mass properties of a hemisphere with a cylindrical hole in the base. The geometry of the object is shown in Fig. 3.31. Figure 3.32 presents the example as it would appear on the screen.

#n# CALCULATE MPROPERTIES ROTATE SAREA MPOINT:

PLEASE DIGITIZE TWO POINTS TO DEFINE ROTATION AXIS
MODEL loc d_1d_2 *[CR]*

PLEASE ENTER SOLID BOUNDARY # 1: $d_3d_4d_5d_6d_7$ *[CR]*
PLEASE ENTER SOLID BOUNDARY # 2: *[CR]*

VOLUME	$= 0.5340707E+02$
MASS	$= 0.5340707E+02$
SURFACE AREA	$= 0.9110618E+02$

CENTROID:
$X = -0.2780463E-06$ $Y = 0.1161765E+01$ $Z = 0.0000000E+00$
FIRST MOMENTS:
$FX = -0.1484964E-04$ $FY = 0.60204645E+02$ $FZ = 0.0000000E+00$
MOMENTS OF INERTIA:
$IXX = 0.2017426E+03$ $IYY = 0.2020044E+03$ $IZZ = 0.2017426E+03$
PRODUCTS OF INERTIA:
$IXY = -0.1529058E-04$ $IXZ = -0.1462013E-10$ $IYZ = 0.0000000E+00$

Example 8

Calculate the mass properties of a cylinder with a square hole along an axis. The geometry of the object is shown in Fig. 3.33. Figure 3.34 presents the example as it would appear on the screen.

#n# CALCULATE MPROPERTIES PROJECT HIBND 2 HOLE SAREA POINT:

PLEASE DIGITIZE TWO POINTS TO DEFINE PROJECTION AXIS
MODEL loc X0Y0Z0, X0Y0Z1

PLEASE ENTER SOLID BOUNDARY # 1: d_1 *[CR]*
PLEASE ENTER SOLID BOUNDARY # 2: *[CR]*
PLEASE ENTER HOLE # 1: d_2 *[CR]*
PLEASE ENTER HOLE # 2: *[CR]*
VOLUME   *= 0.2313274E+02*
MASS   *= 0.2313274E+02*
SURFACE AREA   *= 0.4826548E+02*
CENTROID:
X = 0.0000000E+00   *Y = −0.3145307E-12*   *Z = 0.1000000E+01*
FIRST MOMENTS:
FX = 0.0000000E+00   *FY = −07275958E-11*   *FZ = 0.2313274E+02*
MOMENTS OF INERTIA:
IXX = 0.5580974E+02   *IYY = 0.5580977E+02*   *IZZ = 0.4993220E+02*
PRODUCTS OF INERTIA:
IXY = −0.7275958E-11   *IXZ = 0.0000000E+00*   *IYZ = −0.7275958E-11*

Example 9

Calculate the mass properties of a section of a pyramid with a cylindrical hole. The geometry of the object is shown in Fig. 3.35.
Figure 3.36 presents the example as it would appear on the screen.
The system displays one cross section at a time because in this example
each cross section has been placed on a different layer.

#n# CALCULATE MPROPERTIES EXSECTION LAY 4 ILAY 2 NSTEP
6 HOLE MPOINT SAREA :

CROSS-SECTION # 1

PLEASE ENTER SOLID BOUNDARY # 1: d_1 *[CR]*
PLEASE ENTER SOLID BOUNDARY # 2: *[CR]*
PLEASE ENTER HOLE # 1: d_2 *[CR]*
PLEASE ENTER HOLE # 2: *[CR]*

CROSS-SECTION # 2

PLEASE ENTER SOLID BOUNDARY # 1: d_3 *[CR]*
PLEASE ENTER SOLID BOUNDARY # 2: *[CR]*
PLEASE ENTER HOLE # 1: d_4 *[CR]*
PLEASE ENTER HOLE # 2: *[CR]*

CROSS-SECTION # 3

PLEASE ENTER SOLID BOUNDARY # 1: d_5 *[CR]*
PLEASE ENTER SOLID BOUNDARY # 2: *[CR]*
PLEASE ENTER HOLE # 1: d_6 *[CR]*
PLEASE ENTER HOLE # 2: *[CR]*

CROSS-SECTION # 4

PLEASE ENTER SOLID BOUNDARY # 1: d_7 *[CR]*
PLEASE ENTER SOLID BOUNDARY # 2: [CR]
PLEASE ENTER HOLE # 1: d_8 *[CR]*
PLEASE ENTER HOLE # 2: [CR]

CROSS-SECTION # 5

PLEASE ENTER SOLID BOUNDARY # 1: d_9 *[CR]*
PLEASE ENTER SOLID BOUNDARY # 2: [CR]
PLEASE ENTER HOLE # 1: d_{10} *[CR]*
PLEASE ENTER HOLE # 2: [CR]

CROSS-SECTION # 6

PLEASE ENTER SOLID BOUNDARY # 1: d_{11} *[CR]*
PLEASE ENTER SOLID BOUNDARY # 2: [CR]
PLEASE ENTER HOLE # 1: d_{12} *[CR]*
PLEASE ENTER HOLE # 2: [CR]

VOLUME $= 0.4594673E+02$
MASS $= 0.4594673E+02$
SURFACE AREA = $= 0.8906595E+02$
CENTROID:
$X = 0.1006903E-06$ $Y = 0.7002763E-07$ $Z = 0.2182522E+01$
FIRST MOMENTS:
$FX = 0.4626389E-05$ $FY = -.3217541E-05$ $FZ = 0.1002797E+03$
MOMENTS OF INERTIA:
$IXX = 0.3751889E+03$ $IYY = 0.3751889E+03$ $IZZ = 0.1227238E+03$
PRODUCTS OF INERTIA:
$IXY = 0.9830589E-05$ $IXZ = 0.1314150E-04$ $IYZ = 0.9586305E-05$

Mathematical Solution

(Fig. 3.38)

Volume

Volume of truncated pyramid

$$(V_1) = \int_0^4 \int_0^3 \int_0^5 dz dy \, dx = \int_0^5 \left(\frac{1}{5} z + 3\right)^2 dz$$

$$y = \frac{1}{5} z + 3$$

$$X = \frac{1}{5} z + 3$$

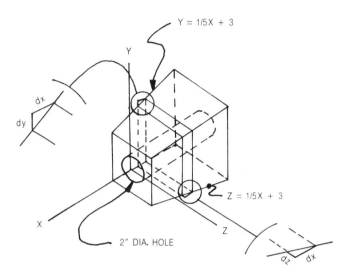

FIGURE 3.38

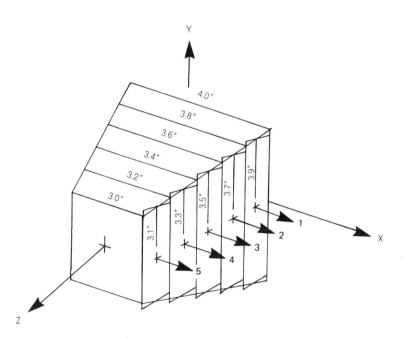

FIGURE 3.39

$$V_1 = \int_0^5 \frac{1}{25} z^2 + \frac{6}{5} z + 9 \; dx = \frac{1}{25} \frac{z^3}{3} + \frac{6}{5} \frac{z^2}{2} + 9 z \Big|_0^5$$

$$= \frac{1}{25} \cdot \frac{125}{3} + \frac{6}{5} \cdot \frac{25}{2} + 9 \cdot 5 = 61.666666667 \text{ in.}^3$$

Volume of hole $(V_2) = \pi r^2 L = \pi (1)^2 (5) = 15.70796327 \text{ in.}^3$

Volume of solid $= V_2 - V_1 = 61.666666667 - 15.70796327$

$$= 45.95870339 \text{ in.}^3$$

$$= \frac{\iiint \rho z \; dz \; dy \; dx}{\iiint \rho \; dz \; dy \; dx}$$

$\rho = 1$, homogeneous materials.

$$z = \frac{\int_0^4 \int_0^3 \int_0^5 x \; dz \; dy \; dx - [V_{hole}(x)]}{45.95870339}$$

$$= \frac{\int_0^5 [1/25)x^3 + (6/5)x^2 + 9x] dx - [15.70796327(2.5)]}{45.95870339}$$

$$= \frac{[(1/25)(x^4/4) + (6/5)(x^3/3) + (9)(x^2/2)]_0^5 - 39.26990818}{45.95870339}$$

$$= \frac{(6.25 + 50 + 112.5) - 39.26990818}{45.95870339} = \frac{129.4800918}{45.95870339}$$

$$= 2.817313855 \text{ in.}$$

or from the opposite side of the base:

$z = 5.0 - 2.817313855 = 2.182686145 \text{ in.}$

Moment of Inertia (Fig. 3.39)

Section 1

$$I_{xx_1} = \frac{1}{12} m(a^2 + b^2) = \frac{1}{12} (3.9)^2 (1)(1^2 + 3.9^2) = 20.55 \text{ in.}^4$$

Section 2

$$I_{xx_2} = \frac{1}{12} (3.7)^2 (1)(1^2 + 3.7^2) = 16.76 \text{ in.}^4$$

Section 3

$$I_{xx_3} = \frac{1}{12} (3.5)^2 (1)(1^2 + 3.5^2) = 13.53 \text{ in.}^4$$

Section 4

$$I_{xx_4} = \frac{1}{12} (3.3)^2 (1)(1^2 + 3.3^2) = 10.79 \text{ in.}^4$$

Section 5

$$I_{xx_5} = \frac{1}{12} (3.1)^2 (1)(1^2 + 3.1^2) = 8.50 \text{ in.}^4$$

Section 1

$$md^2_{xx_1} = (3.9)^2 (1)(0.5)^2 = 3.80 \text{ in.}^4$$

Section 2

$$md^2_{xx_2} = (3.7)^2 (1)(1.5)^2 = 30.80 \text{ in.}^4$$

Section 3

$$md^2_{xx_3} = (3.5)^2 (1)(2.5)^2 = 76.56 \text{ in.}^4$$

Section 4

$$md^2_{xx_4} = (3.3)^2 (1)(3.5)^2 = 133.40 \text{ in.}^4$$

Section 5

$$md^2_{xx_5} = (3.1)^2 (1)(4.5)^2 = 194.60 \text{ in.}^4$$

Total moment of inertia of pyramid about the X axis is 20.55 + 16.76 + 13.53 + 10.79 + 8.50 + 3.804 + 30.80 + 76.56 + 133.40 + 194.60 = 509.29 in.^{4} = I_{xx} = I_{yy}.

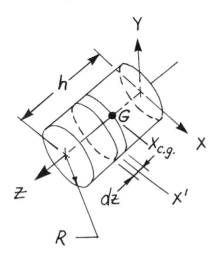

FIGURE 3.40 CALCULATE VOLUME using the XSECTION modifier.

Cylinder

Consider the cylinder (Fig. 3.40) as made up of a series of thin disks of height dz and mass (m dz/h). The moment of inertia of the thin disk about its x' axis parallel to the x axis is given by

$$I_{x'} = \frac{1}{4} \, m \frac{dz}{h} \, R^2$$

Transferring to the x axis by the parallel-axis theorem yields

$$I_{xx_{c.g.}} = I_{x'} + \left(m \frac{dz}{h}\right)z^2 = \frac{1}{4}\left(m \frac{dz}{h}\right)R^2 + \left(m \frac{dz}{h}\right)z^2$$

To determine I_x for the entire cylinder, sum the I_x for all disks.

$$I_{xx_{c.g.}} = \frac{mR^2}{4h} \int_0^h dz + \frac{m}{h} \int_0^h z^2 \, dz = \frac{1}{4} mR^2 + \frac{1}{12} mh^2$$

$$= \frac{1}{12} \, m(3R^2 + h^2)$$

and m = V, since ρ = 1.

$$I_{xx_{c.g.}} = \frac{1}{12}(15.70796327)[3(1)^2 + 5^2]$$

$$= 36.6519143 \text{ in.}^4$$

Transferring to the x axis yields

$$I_{xx} = I_{xx_{c.g.}} + Vd_{xx}^2$$

$$= 36.6519143 + 15.70796327(2.5)^2$$

$$= 36.6519143 + 98.17477044$$

$$I_{yy} = I_{xx} = 134.8266847 \text{ in.}^4$$

This value is a negative moment of mass inertia since it is a hole. Therefore,

$$I_{total} = I_{pyramid} - I_{hole}$$

or

$$I_{xx} = I_{xx_{pyramid}} - I_{xx_{hole}}$$

$$= 509.29 - 134.83$$

$$= 374.463 \text{ in.}^4 = I_{yy}$$

Another Method

$$x = 4 - 0.2z$$

$$y = 4 - 0.2z$$

$$I_{xx}\big|_p = I_{yy}\big|_p = \int_0^5 z^2(4 - 0.2z)^2 dz$$

$$= I_{yy}\big|_p = \frac{16z^3}{3} - \frac{1.6z^4}{4} + \frac{0.04z^5}{5}\Big]_0^5$$

$$= I_{yy}\big|_p = 666.67 - 250 + 25 = 441.67 \text{ in.}^4$$

$$I_{xx}\big|_c = -134.83 \text{ in.}^4$$

$$I_{xx}\big|_T = 441.67 - 134.83 = 306.84 \text{ in.}^4$$

This is a more accurate answer than that obtained using method shown in Fig. 3.38, which was the method used with the ICGS.

Example 10

Calculate the mass properties of a 1 by 1 by 2 box without the use of layer echoing. The geometry of the object is shown in Fig. 3.37.

#n# CALCULATE MPROPERTIES XSECTION NSTEP 3 SAREA
MPOINT:

CROSS-SECTION # 1

PLEASE ENTER SOLID BOUNDARY # 1: d_1 *[CR]*
PLEASE ENTER SOLID BOUNDARY # 2: *[CR]*

CROSS-SECTION # 2

PLEASE ENTER SOLID BOUNDARY # 1: d_2 *[CR]*
PLEASE ENTER SOLID BOUNDARY # 2: *[CR]*

CROSS-SECTION # 3

PLEASE ENTER SOLID BOUNDARY # 1: d_3 *[CR]*
PLEASE ENTER SOLID BOUNDARY # 2: *[CR]*

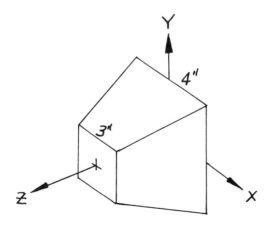

FIGURE 3.41

VOLUME   = 0.2000000E+01
MASS   = 0.2000000E+01
SURFACE AREA   = 0.1000001E+02
CENTROID:
X = 0.5000000E+00   Y = 0.5000000E+00   Z = 0.1000000E+01
FIRST MOMENTS:
FX = 0.1000000E+01   FY = 0.1000000E+01   FZ = 0.2000000E+01
MOMENTS OF INERTIA:
IXX = 0.3333333E+01   IYY = 0.3333333E+01   IZZ = 0.1333333E+01
PRODUCTS OF INERTIA:
IXY = 0.5000000E+00   IXZ = 0.1000000E+01   IYZ = 0.1000000E+01

Mathematical Solution

Volume

V = (area of base)(height) = (1)(1)(2) = 2 in.

Surface area

Surface area = $2[(1)(1)] + 4[(1)(2)] = 2 + 8 = 10$ in.

Centroid

Symmetry of the rectangular box's sides yields $\bar{x} = 0.5$ in., $\bar{y} = 0.5$ in., and $\bar{z} = 1$ in.

Mass moment of inertia

Figure 3.42 shows that the moment of inertia of the block about the z axis is equal to the summation of a series of thin plates each of thickness dz, cross section b by c, and mass dm. First determine I_x for a thin plate with cross section b by c, thickness dz, and mass dm (see Fig. 3.43). Since $I_x = I'_x + I'_y$, find I'_x and I'_y to obtain the result. However, I'_x is really the summation of the centroidal moments of a

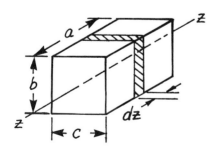

FIGURE 3.42

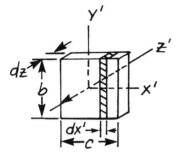

FIGURE 3.43

series of bars of mass dm' and of negligible cross section (dx by dz) and height b. This may then be written

$$I'_x = \int \frac{1}{12} (dm')b^2 = \frac{1}{12} (b^2) \, dm$$

Similar reasoning yields

$$I'_y = \int \frac{1}{12} (dm')c^2 = \frac{1}{12} (c^2) \, dm$$

From this it follows that I_x for a thin plate of mass dm is equal to I'_x + I'_y, or $I_x = (1/12)(dm)(B^2 + c^2)$.

For the entire block it is seen that

$$I_z = \int (1/12)(dm)(b^2 + c^2) = (1/12)m(b^2 + c^2),$$

or more rigorously, since dm = (dz/a)m,

$$I_z = \int_0^a \frac{1}{12} \left(\frac{m}{a}\right)(b^2 + c^2)dz = \frac{1}{12} m (b^2 + c^2)$$

Similarly,

$$I_x = \frac{1}{12} m(a^2 + b^2) \quad \text{and} \quad I_y = \frac{1}{12} m(a^2 + c^2)$$

and if this mass moment of inertia is transferred to the x, y, and z axes, and $\rho = 1$ or $m = V\rho = 1$, then

$$I_x = \frac{1}{12} m(a^2 + b^2) + md_{xx}^2$$

$$I_y = \frac{1}{12} m(a^2 + c^2) + md_{yy}^2$$

$$I_z = \frac{1}{12} m(b^2 + c^2) + md_{zz}^2$$

$$I_x = \frac{1}{12}(2)[2^2 + 1^2] + 2(1.118)^2$$

$$= 0.833 + 2.5000 = 3.333 \text{ in.}^4$$

$$I_y = \frac{1}{12}(2)[1^2 + 1^2] + 2(0.707)^2$$

$$= 0.3333 + 0.9997 = 1.3333 \text{ in.}^4$$

$$I_z = \frac{1}{12}(2)[2^2 + 1^2] + 2(1.118)^2$$

$$= 0.8333 + 2.5000 = 3.3333 \text{ in.}^4$$

Products of mass inertia (Fig. 3.44)

$$I_{xy} = I_{xy} + md_x d_y$$

$$I_{xz} = I_{xz} + md_x d_z$$

$$I_{yz} = I_{yz} + md_y d_z$$

By symmetry it is seen that $I_{xy} = I_{c.g.} = 0$, so that

$$I_{xy} = md_x d_y$$

$$I_{xz} = md_x d_z$$

$$I_{yz} = md_y d_z$$

and if $\rho = 1$, then $m = \rho V = V$, so that

$$I_{xy} = Vd_x d_y = 2(0.5)(0.5) = 0.5000 \text{ in.}^4$$

$$I_{xz} = Vd_x d_z = 2(0.5)(1) - 1.0000 \text{ in.}^4$$

$$I_{yz} = Vd_y d_z = 2(0.5)(1) = 1.0000 \text{ in.}^4$$

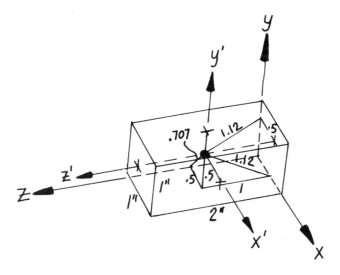

FIGURE 3.44

Example 11

Calculate the mass properties of a two-dimensional washer. Refer to Fig. 3.45, which shows the two-dimensional boundaries and the mass point location.

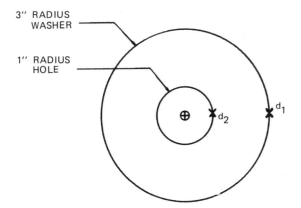

FIGURE 3.45 CALCULATE MPROPERTIES using the PLANE modifier.

#n# CALCULATE MPROPERTIES PLANE HOLE MPOINT :

PLEASE ENTER SOLID BOUNDARY # 1: d_1 [CR]
PLEASE ENTER SOLID BOUNDARY # 2: [CR]
PLEASE ENTER HOLE # 1: d_2 [CR]
PLEASE ENTER HOLE # 2: [CR]

AREA   = 0. 2513274E+02
MASS   = 0. 2513274E+02
CENTROID:
X = −0. 7589099E−06   Y = −0. 4704393E−12   Z = 0. 0000000E+00
FIRST MOMENTS:
FX = −0. 1907349E−04 FY = −0. 1182343E−10   FZ = 0. 0000000E+00
MOMENTS OF INERTIA:
IXX = 0. 6283186E+02 IYY = 0. 6283184E+02   IZZ = 0. 1256637E+03
PRODUCTS OF INERTIA:
IXY = −0. 1818989E−10   IXZ = 0. 0000000E+00   IYZ = 0. 0000000E+00

Mathematical Solution

Area

$$A = \frac{\pi}{4}(d_o^2 - d_i^2) = 0.7853981634(6^2 - 2^2)$$

$$= 25.13274123 \text{ in.}^2$$

Centroid

By symmetry, $\bar{x} = 0$, $\bar{y} = 0$, and $\bar{z} = 0$.

Area moments of inertia

$$I_{xx_1} = I_{yy_1} = 0.0491d^4 = 0.0491(6)^4 = 63.6336 \text{ in.}^4$$

$$I_{xx_2} = I_{yy_2} = 0.0491d^4 = 0.0491(2)^4 = 0.7856 \text{ in.}^4$$

$$I_{xx}|_{total} = I_{xx_1} - I_{xx_2} = 63.6336 - 0.7856$$

$$= 62.848 \text{ in.}^4 = I_{yy}|_{total}$$

$$I_{zz} = 0.0982(d_o^4 - d_i^4) = 2(62.848)$$

$$= 125.695 \text{ in.}^4$$

Products of inertia

Because of the axes of symmetry,

$$I_{xy} = I_{xz} = I_{yz} = 0$$

Notes

1. The XSECTION method for calculating mass properties is approximate because the variation of the shape of the part is not known between cross sections. Therefore, it is advisable to use the PROJECT or ROTATE method whenever possible. If the XSECTION method must be used, it is suggested that more cuts be performed near the region of the part where extreme change of geometry is expected; use as many cuts as possible in that region.
2. The axis for rotation must lie in the plane of the boundaries. The axis for projection must *not* lie in the plane of the boundaries. It is up to you to check that all of the boundaries lie on a single plane and that no boundary intersects itself.
3. If you specify a boundary incorrectly, use the US-DEL key or RUB OUT key to delete the last boundary defined (before another boundary is defined). You can use US-DEL or RUB OUT while defining a boundary to delete the last entity entered. If more deletions are entered than entities, the *entire* previous boundary defined is deleted.
4. When identifying the entities in a boundary, *always* digitize the entities in order (one after another) around a boundary.
5. You should select an empty layer (using the SELECT LAYER command) when using CALCULATE MASS PROPERTIES XSECTION with layer echoing; otherwise, undesired graphics may remain on the screen.

3.3.4 Techniques for Using the Calculate Mass Properties and Calculate Volume Commands

This section contains some suggestions for getting the greatest possible accuracy from the CADDS 4 MPC package using the CALCULATE MPROPERTIES and CALCULATE VOLUME commands. Most important, you should remember that use of the PROJECT or ROTATE modifiers provides exact answers (within the numerical accuracy of the system), while the XSECTION modifier involves an approximation. It is frequently possible to recognize portions of an object for which PROJECT or ROTATE are applicable, and it usually pays to use these modifiers on such portions.

Example 12

The object in Fig. 3.46 consists of a 4-in. cube with a 1-in. radius spherical hole in the center. Although this may appear to be a candi-

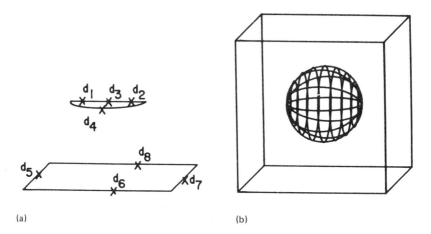

(a)   (b)

FIGURE 3.46 (a) Boundaries that can be projected and rotated to generate a cube with a spherical hole in the center; (b) geometry after projection and rotation of the boundaries.

date for XSECTION because it is neither a solid of projection or a solid of rotation, notice that the cube is a solid of projection and the sphere is a solid of rotation. In the following command printout, the mass properties of each are calculated using PROJECT and ROTATE; then the mass point of the sphere is subtracted (using the MERGE MPOINT command) from that of the cube.

#n# CALCULATE MPROPERTIES ROTATE MPOINT:

PLEASE DIGITIZE TWO POINTS TO DEFINE ROTATION AXIS
MODEL loc END d_1d_2 [CR]

PLEASE ENTER SOLID BOUNDARY # 1: d_3d_4 [CR]
PLEASE ENTER SOLID BOUNDARY # 2: [CR]

VOLUME   = 0.4188789+01
MASS   = 0.4188789E+01
CENTROID:
X = 0.2000000E+01   Y = 0.2000000E+01   Z = 0.2000000E+01
*   FIRST MOMENTS:*
FX = 0.8377578E+01   FY = 0.8377578E+01   FZ = 0.8377578E+01
*   MOMENTS OF INERTIA:*
IXX = 0.3518583E+02   IYY = 0.3518580E+02   IZZ = 0.3518583E+02
*   PRODUCTS OF INERTIA:*
IXY = 0.1675516E+02   IXZ = 0.1675516E+02   IYZ = 0.1675516E+02

#n# CALCULATE MPROPERTIES PROJECT HIBND 4 MPOINT:

PLEASE DIGITIZE TWO POINTS TO DEFINE PROJECTION AXIS
MODEL loc X0Y0Z0, X0Y0Z1 [CR]

PLEASE ENTER SOLID BOUNDARY # 1: $d_5d_6d_7d_8$ *[CR]*
PLEASE ENTER SOLID BOUNDARY # 2: *[CR]*

VOLUME $\quad\quad\quad\quad = 0.6400000E+02$
MASS $\quad\quad\quad\quad\quad\, = 0.6400000E+02$
CENTROID:
$X = 0.2000000E+01 \quad\quad Y = 0.2000000E+01 \quad\quad Z = 0.2000000E+01$
FIRST MOMENTS:
$FX = 0.1280000E+03 \quad\quad FY = 0.1280000E+03 \quad\quad FZ = 0.1280000E+03$
MOMENTS OF INERTIA:
$IXX = 0.6826667E+03 \quad IYY = 0.6826667E+03 \quad IZZ = 0.6826667E+03$
PRODUCTS OF INERTIA:
$IXY = 0.2560000E+03 \quad IXZ = 0.2560000E+03 \quad IYZ = 0.2650000E+03$

#n# MERGE MPOINT SUBTRACT: *MODEL ent* d_9 *MODEL ent* d_{10} *[CR]*

#n# LIST MPOINT: *MODEL ent* d_{11} *[CR]*

VALUES ARE GIVEN WITH RESPECT TO MODEL SPACE COORDIN-
ATES.

VOLUME $\quad\quad\quad\quad = 0.5981121E+02$
MASS $\quad\quad\quad\quad\quad\, = 0.5981121E+02$
CENTROID:
$X = 0.2000000E01 \quad\quad\, Y = 0.2000000E+01 \quad\quad Z = 0.2000000E+01$
FIRST MOMENTS:
$FX = 0.1196223E+03 \quad\quad FY = 0.119622E+03 \quad\quad FZ = 0.1196224E+03$
MOMENTS OF INERTIA:
$IXX = 0.6474809E+03 \quad IYY = 0.6474809E+03 \quad IZZ = 0.6474809E+03$
PRODUCTS OF INERTIA:
$IXY = -.2392448E+03 \quad IXZ = 0.2392448E+03 \quad IYZ = 0.2392448E+03$

Mathematical Solution (Fig. 3.47)

Volume

V_1 = volume of cube = a(b)c = 4(4)(4) = 64 in.^{3}

V_2 = volume of sphere + $\frac{4}{3}\pi r^3 = \frac{4}{3}\pi(1) = 4.188790205$ in.^{3}

$V_{total} = V_1 - V_2 = 64 - 4.188790205 = 59.8112098$ in.^{3}

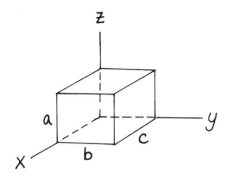

FIGURE 3.47

Centroid

Because of symmetry, x = 2 in., y = 2 in., and z = 2 in.

Mass moments of inertia

Sphere, where $\rho = 1$:

$$I_{xx}\big|_{c.g.} = I_{yy}\big|_{c.g.} = I_{zz}\big|_{c.g.} = \frac{2}{5} VR^2$$

$$= \frac{2}{5}(4.188790205)(1)^2 = 1.675516082 \text{ in.}^4$$

$$I_{xx_1} = I_{yy_2} = I_{zz_3}$$

$$= I_{xx}\big|_{c.g.} + Vd^2 = 1.675516082 + 4.188790205(\sqrt{8})^2$$

$$= 35.18583772 \text{ in.}^4 \qquad \text{this value is negative}$$

Cube, where $\rho = 1$:

$$I_{xx_2} = I_{yy_2} = I_{zz_2} = \frac{2Va^2}{3} = \frac{2(64)(4)^2}{3} = 682.6666667 \text{ in.}^4$$

$$I_{xx}\big|_T = I_{xx_2} - I_{xx_1} = 682.6666667 - 35.18583772$$

$$I_{xx}\big|_T = I_{yy}\big|_T = 647.480829 \text{ in.}^4$$

Products of mass inertia

Because the coordinate planes are all planes of symmetry for the mass distribution, the products of inertia are zero or $I_{xy} = I_{xz} = I_{yz} = 0$, and the products of inertia for the sphere are

$$I_{xy} = \bar{I}_{xy} + md_x d_y$$

$$I_{xz} = \bar{I}_{xz} + md_x d_z$$

$$I_{yz} = \bar{I}_{yz} + md_y d_z$$

If $\rho = 1$ and $m = \rho V$, then $m = V$ and

$$I_{xy} = 0 + Vd_x d_y = (4.188789)(2)(2)$$

$$= 16.755156 \text{ in.}^4$$

$$I_{xz} = 0 + Vd_x d_z = (4.188789)(2)(2)$$

$$= 16.755156 \text{ in.}^4$$

$$I_{yz} = 0 + Vd_y d_z = (4.188789)(2)(2)$$

$$= 16.755156 \text{ in.}^4$$

The values for I_{xy}, I_{xz}, and I_{yz} are negative.

Products of inertia for the cube

$$I_{xy} = \int_0^a \int_0^b \int_0^c xy\rho \ dx \ dz = \int_0^a \int_0^b \frac{c^2}{c} y\rho \ dy \ dz$$

$$= \frac{c^2 b^2}{4} \rho \ dz = ac^2 \frac{b^2}{4} \rho = \frac{\rho V}{4} cb$$

$$I_{xz} = \frac{\rho V}{4} ac$$

$$I_{yz} = \frac{\rho V}{4} ab$$

If $\rho = 1$, then

$$I_{xy} = \frac{V}{4} (4)(4) = 64(4) = 256 \text{ in.}^4$$

$$I_{xz} = \frac{V}{4} (4)(4) = 64(4) = 256 \text{ in.}^4$$

$$I_{yz} = \frac{V}{4} (4)(4) = 64(4) = 256 \text{ in.}^4$$

Total product of inertia

$$I_{xy} = 256 - 16.755156 = 239.244844 \text{ in.}^4$$

$$I_{xz} = 256 - 16.755156 = 239.244844 \text{ in.}^4$$

$$I_{yz} = 256 - 16.755156 = 239.244844 \text{ in.}^4$$

If you decide that XSECTION is the best alternative, there are several things to keep in mind for the best results. First, the XSECTION method is designed to approximate the mass properties of an object whose cross section does not change much from one to the next. If the geometry changes quickly over some region of the object, you should include many cross sections from that region. If the object has a sudden jump in the area or shape of the cross sections at some point (like the object in Fig. 3.48), you will get better results by doing the calculation in two steps: (1) calculate up to the jump, and (2) calculate from there onward. Then you can merge the mass points to find the mass properties of the whole object.

Second, the accuracy of results obtained with the XSECTION modifier depends on the number of cuts (cross sections) used. To illustrate this effect, mass properties are calculated on the object shown in Fig. 3.27. The object is defined by a ruled surface between two ellipses on parallel planes 10 units apart, with their major axes perpendicular to each other. The ellipse with its center at X0Y0Z0 has a half major

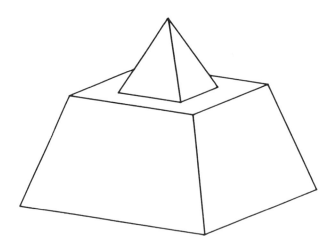

FIGURE 3.48 Object with a step change in geometry.

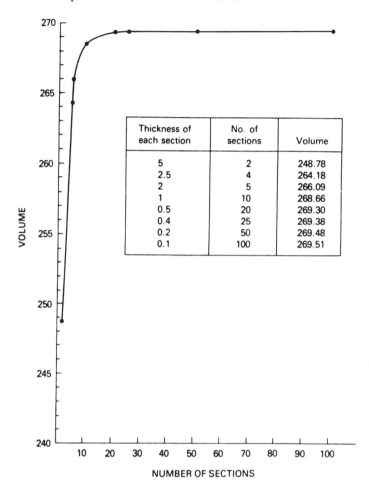

Thickness of each section	No. of sections	Volume
5	2	248.78
2.5	4	264.18
2	5	266.09
1	10	268.66
0.5	20	269.30
0.4	25	269.38
0.2	50	269.48
0.1	100	269.51

FIGURE 3.49 Calculated volume of the object shown in Fig. 3.27 plotted against the number of cuts used in the calculation. Number of steps = number of sections + 1.

axis of 5 and a half minor axis of 2. The ellipse with its center at X0Y0Z0 has a half major axis of 5 and a half minor axis of 1.5. It is cut with a plane perpendicular to the Z axis into different numbers of sections. Mass property calculations are performed on the cuts.

Figure 3.49 is a graph of the calculated volume versus the number of cuts used. It shows that for this geometry, increased accuracy is achieved as the number of cuts is increased to about 20, beyond which more cuts do not improve the answer significantly. A useful technique

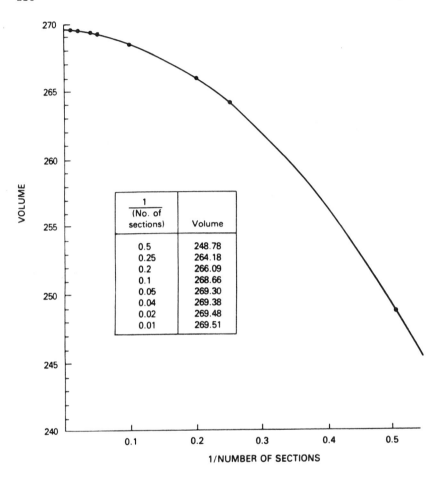

$\dfrac{1}{(\text{No. of sections})}$	Volume
0.5	248.78
0.25	264.18
0.2	266.09
0.1	268.66
0.05	269.30
0.04	269.38
0.02	269.48
0.01	269.51

FIGURE 3.50 Calculated volume of the object shown in Fig. 3.27 plotted against the inverse of the number of cuts used in the calculations. Number of steps = number of sections + 1.

for extrapolating the correct answer is to plot there data versus the reciprocal of the number of cuts, as shown in Fig. 3.50. You can see that the volume of the object is between 269 and 270.

As a further illustration, the graphical technique in Fig. 3.49 is repeated for the geometry shown in Fig. 3.35. The volumes are plotted and listed in Fig. 3.51. The convergence of the volume to the true answer (45.9587) is seen as more cuts are used.

Actually, this geometry is an example of the situation discussed at the beginning of this section, where XSECTION alone is not the best

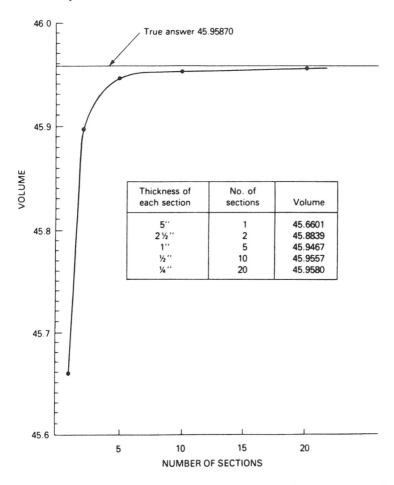

The figure contains the following table:

Thickness of each section	No. of sections	Volume
5″	1	45.6601
2½″	2	45.8839
1″	5	45.9467
½″	10	45.9557
¼″	20	45.9580

True answer 45.95870

FIGURE 3.51 Calculated volume of object in Fig. 3.35 plotted against number of cuts used in the calculation. Number of steps = number of sections + 1.

technique to use. Because the cylindrical hole in the center of the frustum of the pyramid is a tabulated cylinder, you can obtain more accurate results by using the XSECTION method on the frustum of the pyramid and the PROJECT method on the cylindrical hole. Afterward, you can find the mass properties of the complete object by merging (subtracting) the mass point of the cylinder and the mass point of the frustum of the pyramid.

With 10 sections (11 cuts) of the frustum of the pyramid and use of the PROJECT method on the cylindrical hole, the volume is calculated as equal to the true answer (45.9897). However, it is important to understand that the true volume is found because the formulas employed in finding the *volume* in the XSECTION method are geared toward a more exact solution for geometries which are similar to the frustum of a pyramid.

4
Stress and Strain

4.1 STRESS

Stress is force divided by area, in the limit as the area approaches zero. For example, Fig. 4.1a shows a force dF which is exerted on an area dA on the end of a rectangular bar. The stress is $\sigma = dF/dA$. Figure 4.1b shows the distribution of force over the entire end of the bar. Here the total is the sum of the forces dF, and the total area is the sum of the areas dA. Stress is usually expressed in pounds per square inch, which is often abbreviated as psi. Stress is a vector.

Normal stress is stress for which the vector is normal, or perpendicular, to the area on which it acts. The stress represented in Fig. 4.1a is a normal stress. Normal stress is represented here by σ (sigma).

A normal stress is called a *tensile stress* (or tension) when it stretches the material on which it acts, as in Fig. 4.2a. A tensile stress is usually taken to be positive.

A normal stress is called a *compressive stress* (or compressive) when it shortens the material on which it acts, as in Fig. 4.2b. A compressive stress is usually taken to be negative.

Shear stress is stress for which the vector is tangent to the area on which it acts. Figure 4.3 shows a shear stress acting on the end of a bar. Shear stress is represented here by τ (tau).

4.2 STRAIN

Strain is deformation divided by the length in which the deformation occurs. Since strain is a length divided by a length, it is dimensionless. Strain is a vector.

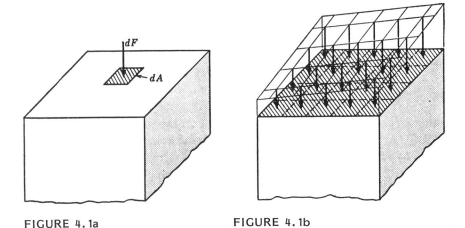

FIGURE 4.1a   FIGURE 4.1b

Normal strain is deformation divided by a length which is parallel to the deformation. Normal strain is represented here by ε (epsilon).

Figure 4.4a shows a body that is subjected to the forces F_1, F_2, F_3, and F_4, and a small element with dimensions dx and dy. Figure 4.4b shows the element subjected to normal strain; the rectangle ABCD is the element before straining; the rectangle $AB_1C_1D_1$ is the element after straining.

The deformation in the x direction is a dx, and the normal strain in the x direction is

$$\varepsilon_x = \frac{adx}{dx} = a$$

which is also called the x component of the strain. The deformation in the y direction is b dy; therefore, the normal strain in the y direction is

$$\varepsilon_y = \frac{bdy}{dy} = b$$

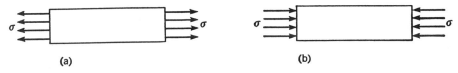

(a)   (b)

FIGURE 4.2 (a) Tensile stress; (b) compressive stress.

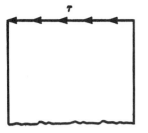

FIGURE 4.3 Shear stress.

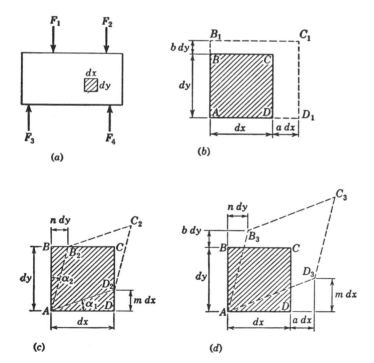

FIGURE 4.4 (a) Body; (b) element subjected to normal strain; (c) element subjected to shear strain; (d) element subjected to general condition of strain.

which is also called the y component of the strain.

Shear strain is deformation divided by a length which is perpendicular to the deformation. Shear strain is represented here by γ (gamma).

Figure 4.4c shows the element subjected to shear strain. The rectangle ABCD is the element before straining; the rhombus $AB_2C_2D_2$ is the element after straining.

The shear strain is defined as the change in the right angle at A; but with the condition that the deformation is very small; it is satisfactory to take the change in the right angle at A as the sum of the tangents of the angles α_1 and α_2. Thus the shear strain is

$$\gamma = \tan \alpha_1 + \tan \alpha_2 = \frac{mdx}{dx} + \frac{n \, dy}{dy} = m + n$$

Figure 4.4d shows the element in a general condition of strain, the result of superposing Fig. 4.4b and c. Here the rectangle ABCD is the element before straining, and the rhombus $AB_3C_3D_3$ is the element after straining.

4.3 HOOKE'S LAW

Hooke's law is a relation between stresses and strains which has been deduced from experimental results. It is approximately true for some materials, chiefly metals, as long as the stresses do not exceed certain limits.

Figure 4.5 shows a rectangle ABCD before being subjected to stress and the same rectangle $AB_1C_1D_1$ after being subjected to stress. Strain occurs when the rectangle is stressed. Hooke's law gives the relation between the stresses and the strains as

$$\varepsilon_x = \frac{1}{E} (\sigma_x - \nu\sigma_y)$$

$$\varepsilon_y = \frac{1}{E} (\sigma_y - \nu\sigma_x)$$

Here there are two elastic constants of the material:

1. The modulus of elasticity E, usually expressed in pounds per square inch (psi)
2. Poisson's ratio ν (nu), which is dimensionless.

The modulus of elasticity represents the stiffness of the material, that is, its resistance to deformation. Poisson's ratio represents a lateral

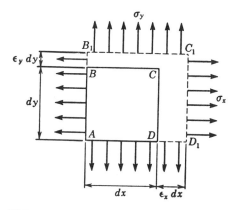

FIGURE 4.5

effect, the fact that a stress in one direction is accompanied by a strain in the perpendicular direction.

Figure 4.6 shows a rectangle ABCD in an unstressed condition and a parallelogram $AB_1C_1D_1$ which the rectangle becomes after being subjected to the shear stress τ_{xy}. The relation between shear stress and shear strain is

$$\gamma_{xy} = \frac{\tau_{xy}}{G} \tag{1}$$

where G is the shear modulus and is usually expressed in pounds per square inch. The shear modulus is a third elastic constant of the material; however, it is not independent of E and ν, and it is possible to show that it is related to them by

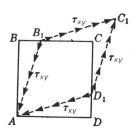

FIGURE 4.6

$$G = \frac{E}{2(1 + \nu)} \qquad (2)$$

4.4 UNIFORM AXIAL STRESS

The simplest stress distribution is uniform tension or compression in the direction of the axis of a bar. If the axis chosen is labeled the x axis, the stress σ_x is constant. Thus

$\sigma_x = $ constant

$\sigma_y = 0$

$\tau_{xy} = 0$

and this set of stresses satisfies the equations of equilibrium,

$$\frac{\delta \sigma_x}{\delta x} + \frac{\delta \tau_{xy}}{\delta y} = 0 \qquad \frac{\delta \sigma_y}{\delta y} + \frac{\delta \tau_{xy}}{\delta x} = Y$$

if the body force Y is zero.

Figure 4.7a shows a uniform tensile stress acting at the end cross sections of the center part of a bar. Since the stress is uniformly distributed over each cross section of the bar, the resultant force must act through the centroid of the bar; this resultant force is an axial force. Figure 4.7b shows the axial forces P acting on the ends of the bar.

The force P is equal to the product of the stress σ and the area A of the cross section of the bar. Thus

$P = A\sigma$

(a)

(b)

FIGURE 4.7

or, in another form often used,

$$\sigma = \frac{P}{A}$$

A limitation on these equations is that the force P must act through the centroid of the cross section of the bar.

4.5 AXIAL STRAIN AND DEFORMATION

When the only stress is a uniform axial stress σ_x, Hooke's law reduces to

$$\varepsilon_x = \frac{\sigma_x}{E}$$

$$\varepsilon_y = - \frac{\nu \sigma_x}{E}$$

A bar in tension undergoes an elongation in the direction in the perpendicular direction, as shown in Fig. 4.8. Here the elongation of the bar is

$$e = \varepsilon_x L$$

and the contraction is

$$c = \varepsilon_y b$$

The equation

$$\sigma = \frac{P}{A}$$

is often used when the axial stress is not uniform but varies along a bar. An example is a tapered bar subjected to an axial force; both

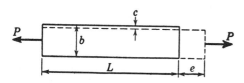

FIGURE 4.8

stress and strain vary along the bar. The results are not exactly correct, but the error is small if the taper of the bar is gradual.

4.6 UNIFORM SHEAR STRESS

There are many problems in which it is of advantage to assume a uniform shear stress. For example, Fig. 4.9a shows two bars which are fastened together by a bolt. Failure may occur by shearing the bolt at sections between the two bars. Figure 4.9b shows the free-body diagram of the center bar and a portion of the bolt; the force P is balanced by shear stress τ on the cross sections of the bolt. The shear stress is assumed to be constant and equal to

$$\tau = \frac{P}{A}$$

where A is the total area in shear. This formula is probably never exactly correct, but it is widely used, since an exact analysis would be very difficult in most instances of its application.

4.7 WARPING OF CROSS SECTION

Usually, the cross section of a bar in torsion warps; that is, the cross section does not remain plane. The easiest way to show that the cross section warps is to assume that it does not warp and then show that this is not in agreement with the requirement for the direction of the shear stress at the boundary, except in one special case.

Figure 4.10a shows a rectangular bar which is subjected to a torque T; the right-hand end rotates (about an axis through point O) through the angle θ with respect to the left-hand end. The point C_1 moves to point C_2 in a direction perpendicular to the line OC_1. If the cross section remains plane, the shear strain is

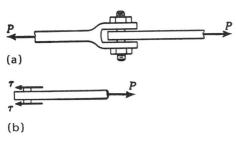

(a)

(b)

FIGURE 4.9

FIGURE 4.10a

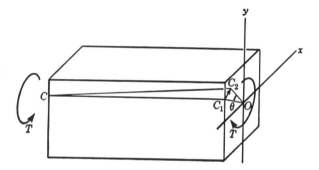

FIGURE 4.10b

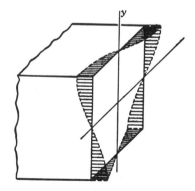

FIGURE 4.10c

$$\gamma = \frac{C_1 C_2}{CC_1}$$

in the direction $C_1 C_2$; and the shear stress must be in this direction. However, the shear stress cannot be in the direction $C_1 C_2$ because the shear stress must be in the direction of the boundary, that is, in a vertical direction. The cross section must warp in such a manner that there is no shear strain in a horizontal direction at C_1 because there is no shear stress in a horizontal direction at C_1. Figure 4.10b shows a view looking downward at the horizontal plane through C and C_1. As the line CC_1 changes direction to CC_2, the cross section must warp, as shown by the dashed line, so that which are originally perpendicular remain perpendicular and there is no shear strain in the horizontal plane.

Figure 4.10c illustrates the warping of a rectangular cross section. In the very special case in which the tangent at every point on the boundary is perpendicular to a line that passes through the center of the cross section, the cross section does not warp. The only boundary that meets this requirements is a complete circle.

4.8 SPECIAL CASE OF A CIRCULAR CROSS SECTION

When the cross section of a bar in torsion is circular, the cross section remains plane. Then a point C_1 in Fig. 4.11a moves to C_2, a distance of $r\theta$, where r is the distance of C_1 from the center of the circle and is the angle with whicht he right-hand cross section rotates with respect to the left-hand cross section. In Fig. 4.11b, the shear strain is

$$\gamma = \frac{C_1 C_2}{CC_1} = \frac{r\theta}{L}$$

As long as the stress does not exceed the proportional limit,

$$\tau = G\gamma = \frac{G\theta}{L} r \tag{3}$$

where G is the shear modulus of the material.

Figure 4.12 shows a hollow circular cross section of inner diameter d_1 and outer diameter d_2. The torque T is the resultant of the shear stress τ. Then, using the center of the circle as a moment center, we have

$$T = \int (\tau \, dA)r = \frac{G\theta}{L} \int r^2 \, dA$$

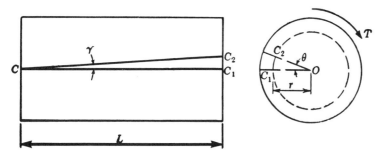

FIGURE 4.11a   FIGURE 4.11b

Here the final integral is the polar moment of inertia of the cross section; that is, it is polar moment of inertia of the larger circular area minus the polar moment of inertia of the smaller circular area. It can be written as

$$r^2 \, dA = J_2 - J_1 = \frac{\pi d_1^4}{32} - \frac{\pi d_1^4}{32}$$

Then

$$T = \frac{G\theta}{L}(J_2 - J_1)$$

$$\theta = \frac{TL}{G(J_2 - J_1)} \tag{4}$$

and θ is called the angle of twist.

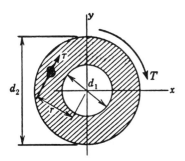

FIGURE 4.12

The maximum shear stress occurs where r is greatest because the shear stress is proportional to r. Then, in Eq. (3),

$$\tau_{max} = \frac{G\theta}{L} \frac{d_2}{2}$$

$$\tau_{max} = \frac{T(d_2/2)}{J_2 - J_1}$$

In applying this theory, it is customary to express quantities in the following units:

T	inch-pounds (in.-lb)
L	inches (in.)
d_1, d_2	inches (in.)
θ	radians (rad)
G	pounds per square inch (psi)

4.9 SHEAR FLOW: SHEAR STRESS IN THIN-WALLED CLOSED SECTIONS

The very elegant concept of shear flow can be established by considering a thin-walled closed section in torsion. Figure 4.13a shows shear stresses τ_1 and τ_2 on such a section. Since the shear stress at a point on a boundary must be in the direction of the tangent at the boundary, and also, since the two boundaries of the thin wall are close together, it is not unreasonable to say that the shear stress is constant in direc-

FIGURE 4.13a

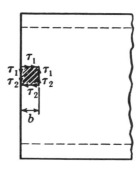

FIGURE 4.13b

tion and magnitude through the wall thickness. This leads to the shear flow, which is the shear force per unit of length, as

$$Q = \tau t$$

where t is the wall thickness.

Next, looking at the view in Fig. 4.13b, the shear stress τ_1 on the horizontal plane must equal the shear stress τ_1 on the vertical plane. Also, the shear stress τ_2 on the horizontal plane must equal the shear stress τ_2 on the vertical plane. The sum of the horizontal forces on the shaded rectangle in Fig. 4.13b must be zero; that is,

$$Q = \tau t$$

or the shear flow Q is constant.

Figure 4.14 shows a thin-walled closed section in torsion. The shear flow is

$$Q = \tau t$$

and the shear force in the length ds is Q ds. The moment of this force with respect to point O is

$$Q \, ds(h \cos \theta) = Q(h \, ds \cos \theta)$$

but since h ds cos is twice the area of the shaded triangle, this moment can be written as 2Q dA.

The torque is the resultant of the shear forces on the cross section, and the torque is a couple and has the same moment with respect to all points. Therefore, the torque is the sum of the moments 2Q dA. Since Q is constant, the torque is

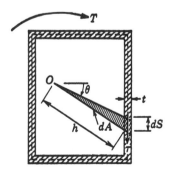

FIGURE 4.14

$$T = 2Q \int dA = 2QA$$

that is, the torque is twice the shear flow times the area enclosed by the midline of the wall. The shear stress is

$$\tau = \frac{Q}{t}$$

so the maximum shear stress occurs at the thinnest point of the wall.

4.10 FLEXURAL STRESS IN STRAIGHT BARS

Flexure is bending. The stress distribution

$$\sigma_x = C_1 y$$

$$\sigma_y = 0$$

$$\tau_{xy} = 0$$

satisfies the equations of equilibrium for stresses when there are no body forces. It can be shown that this stress distribution causes flexure of a straight bar.

Fgiure 4.15a shows this stress distribution on the ends of a straight bar which has the cross section shown in Fig. 4.15b. Figure 4.15c shows a couple M acting on each end of the bar, where M is the resultant of the stresses on the cross section. Since the resultant is a couple, the axial force on the cross section is zero. The axial force is the sum of the forces on the cross section. Therefore,

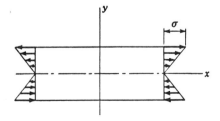

FIGURE 4.15a

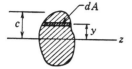

FIGURE 4.15b

$$P = \int \sigma_x \, dA = C_1 \int y \, dA = 0$$

$$y \, dA = 0$$

which shows that the z axis in the cross section passes through the centroid of the cross section.

The couple M is the resultant of the stresses on the cross section. Taking the z axis as a moment axis yields

$$M = (\sigma_x \, dA)y = C_1 \quad y^2 \, dA$$

The last integral is the moment of inertia of the area of the cross section with respect to the z axis, so that

$$M = C_1 I$$

where I is this moment of inertia. The z axis is called the *neutral axis* of the cross section. The surface that contains the neutral axes of all cross sections is called the *neutral surface*.

Since the stress varies directly as the distance from the neutral axis, the maximum stress, σ, occurs at the greatest distance c from the neutral axis. At the distance c,

$$C_1 c = \sigma \qquad C_1 = \frac{\sigma}{c}$$

FIGURE 4.15c

and

$$M = \frac{\sigma I}{c}$$

This is the *flexure formula*, a very widely used formula in engineering and design. It is often written as

$$\sigma = \frac{Mc}{I}$$

or

$$\frac{I}{c} = \frac{M}{\sigma}$$

in which the quantity I/c is called the section modulus. It is custom-ary to take the moment M in inch-pounds, the distance c in inches, and the moment of inertia I in inches to the fourth power in applying the flexure formula.

4.11 ELASTIC CURVE OF A BEAM

Figure 4.16a shows a beam that is caused to bend by the force F. The line AA, which represents the neutral surface, is straight before the load is applied and is a curved line after the load is applied. This line AA is called the *elastic curve* of the beam. The next problem is to find the equation of the elastic curve, from which ordinates and slopes of the curve can be found.

A second y axis is introduced in Fig. 4.16a. The first y axis is an axis in the plane of a cross section of a bar; a y coordinate is the or-dinate of a point in the cross section. The second y axis is used to specify the elastic cuve of a bar; a y coordinate is the ordinate of a point on the elastic curve.

Figure 4.16b shows the elastic curve again, with a particular atten-tion given to the element of length dx. It has already been shown that

$$\frac{d\theta}{dx} = \frac{M}{EI}$$

On the assumption that $d\theta$ is small, it is reasonable to replace $d\theta$ by tan θ, which is equal to dy/dx. Then

$$\frac{d\theta}{dx} = \frac{d}{dx}\left(\frac{dy}{dx}\right) = \frac{d^2y}{dx^2}$$

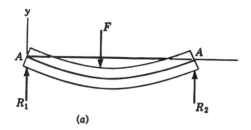

(a)

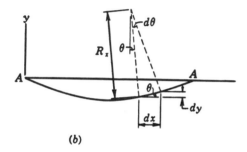

(b)

FIGURE 4.16 (a) Bent beam; (b) elastic curve.

and

$$\frac{d^2 y}{dx^2} = \frac{M}{EI}$$

This is the differential equation from which the equation of the elastic curve can be found. The product EI is the flexure stiffness of the beam and may be constant or variable.

Since the differential equation is of the second order, there are two constants of integration in the resulting equation of the elastic curve. These constants of integration are found by substituting known values of slopes and ordinates in the equation of the elastic curve. These known values of slopes and ordinates are called *boundary conditions*.

4.12 RELATIONS BETWEEN LOAD, SHEAR, AND MOMENT

Figure 4.17 shows a beam, with special attention to an element of length dx, on which are shown the load w, the shear V, and the mo-

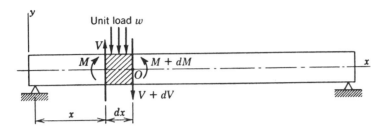

FIGURE 4.17

ment M. The element is in equilibrium; therefore, the sum of the vertical forces on it must be zero. Thus

$$\Sigma F_y = 0$$

$$V - wdx - (V + dV) = 0 \qquad w = - \frac{dV}{dx}$$

which is the relation between the load and the shear. The load is equal to minus the derivative of the shear with respect to the lengthwise coordinate of the beam.

The equilibrium equation for the sum of the moments with respect to point O is

$$\Sigma M_o = 0$$

$$M + V \, dx - w \, dx \, \frac{dx}{2} - (M + dM) = 0$$

in which the third term may be neglected because it is small compared with the other terms. Then

$$V = \frac{dM}{dx}$$

which is the relation between shear and moment.

The load w is shown as a force exerted on the top of the beam in Fig. 4.17. However, the results obtained here would be exactly the same if w were applied at the lower surface of the beam, or somewhere between the top and bottom. If a body force Y is to be taken into account, w must include a term

$$\int Y \, da$$

where the integration is over the cross section of the beam. If the body force is constant,

$$\int Y \, dA = YA$$

4.13 SHEAR STRESS IN A BEAM OF ANY CROSS SECTION

The shear stress at a point on the boundary of a cross section must be in the direction of the tangent to the boundary at that point, as in the circular cross section shown in Fig. 4.18a. The shear stress has a z component as well as a y component on such a line as AA. This presents a three-dimensional stress distribution, the exact analysis of which is beyond the scope of this book. The cross section in Fig. 4.18b also presents a three-dimensional distribution because the shear stress must, in effect, turn the corner at the place where the width of the cross section changes.

A method that gives the average shear stress on a horizontal plane in a beam can be developed by considering the equilibrium of the shaded portion of the beam in Fig. 4.19a. The flexural compressive stress σ on the left-hand side of this block is My_1/I at a distance y_1 from the neutral surface. On the right-hand side of the element, the compressive stress is $(M + dM)y_1/I$, since the moment is $M + dM$ at this cross section. These stresses act on the shaded part of the cross section in Fig. 4.19b. The difference in the total forces that result from these stresses must be balanced by the shear stress τ on the plane AA in Fig. 4.19a. The equilibrium equation for the sum of the horizontal forces gives

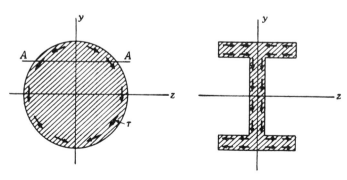

FIGURE 4.18a   FIGURE 4.18b

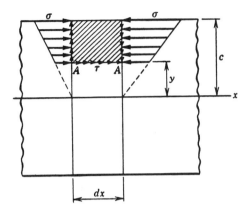

FIGURE 4.19a

$$\sum F_x = 0$$

$$\tau \, dx \, t + \frac{M}{I} \int_y^c y_1 \, dA - \frac{M + dM}{I} \int_y^c y_1 \, dA = 0$$

and

$$\tau = \frac{1}{It} \frac{dM}{dx} \int_y^c y_1 \, dA$$

Then, since $dM/dx = V$,

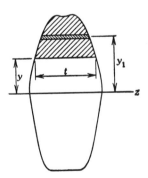

FIGURE 4.19b

$$\tau = \frac{V}{It} \int_y^c y_1 \, dA$$

or

$$\tau = \frac{VQ}{It}$$

where

$$Q = \int_y^c y_1 \, dA$$

Here V is the shear force on the cross section, I is the moment of inertia of the area of the entire cross section, t is the width of the cross section at the plane on which it is desired to find the shear stress, and Q is the moment (with respect to the neutral axis) of that part of the cross section above the plane. It is often possible to calculate this moment of area by a method of composite parts.

4.14 AXIAL STRESS SUPERPOSED ON FLEXURE

The axial stress and the flexural stress are both normal stresses on a transverse cross section of a bar. The two stresses are collinear and can be added algebraically. For example, Fig. 4.20a shows a bar that is subjected to the stress patterns

$$\sigma_1 = \frac{P}{A} \quad \text{and} \quad \sigma_2 = \frac{-My}{I}$$

The final stress pattern is

$$\sigma = \sigma_1 + \sigma_2 = \frac{P}{A} - \frac{My}{I}$$

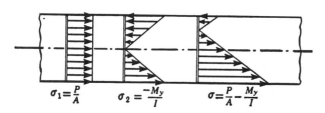

FIGURE 4.20a

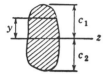

FIGURE 4.20b

Figure 4.20b shows the cross section of the bar. At the top of the bar the normal stress is

$$\sigma = \frac{P}{A} - \frac{Mc_1}{I}$$

and at the bottom of the bar the normal stress is

$$\sigma = \frac{P}{A} + \frac{Mc_2}{I}$$

Here a tensile stress is considered positive and a compressive stress negative. The greatest absolute value of stress occurs where the two stresses are of the same algebraic sign, at the top or bottom of the bar. This stress is usually called the *maximum stress*.

Figure 4.21a shows the stress resultants for an axial stress pattern superposed on a flexural stress pattern; the axial force P is superposed on the bending moment M. The force and couple on the cross section of Fig. 4.21a can be replaced by the eccentric force P of Fig. 4.21b, where the eccentricity is e. Therefore, the combination of axial stress and flexural stress occurs when an eccentric force is applied to a bar. The bending moment on a cross section in Figs. 4.21a and 4.21b is

M = Pe

so that

$$\sigma = \frac{P}{A} \pm \frac{Pey}{I}$$

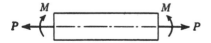

FIGURE 4.21a

FIGURE 4.21b

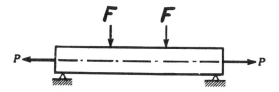

FIGURE 4.22

The combination of axial stress and flexural stress also occurs when a bar is subjected to simultaneous axial and lateral forces, as shown in Fig. 4.22.

4.15 FLEXURE SUPERPOSED ON FLEXURE: GENERAL BENDING

The superposition of one flexural stress pattern on another gives the general case of flexure in a straight bar. Figure 4.23a shows a flexural stress pattern

$$\sigma_x = -\frac{M_z}{I_z} y$$

of the type already considered; Fig. 4.23b shows the z axis in the cross section of the bar. Here the moment is M_z, with respect to the

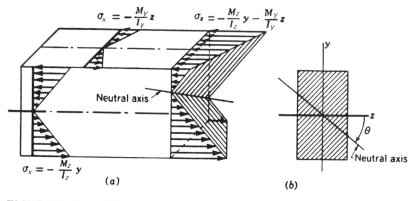

FIGURE 4.23 (a) Stress patterns; (b) cross section.

z axis, and the moment of inertia I_z is with respect to the z axis. A
second flexural stress pattern,

$$\sigma_x = - \frac{M_y}{I_y} z$$

is shown, where the moment M_y and the moment of inertia I_y are with
respect to the y axis. When these two flexural stress patterns are
superposed, the resulting stress pattern is

$$\sigma_x = - \frac{M_z}{I_z} y - \frac{M_y}{I_y} z$$

as shown. The neutral axis of the bar is the line along which σ is
zero. Thus

$$-\frac{M_z}{I_z} y - \frac{M_y}{I_y} z = - \qquad y = -\frac{M_y}{M_z} \frac{I_z}{I_y} z$$

is the equation of the neutral axis in the plane of the cross section.
The inclination of the neutral axis is given by

$$\tan \theta = \frac{y}{z} = - \frac{M_y}{M_z} \frac{I_z}{I_y}$$

The angle θ is negative (clockwise) for the case shown.

It is necessary that the y and z axes be principal axes of the cross
section for this theory to apply, since the simple flexure formula is
valid only for the case of a principal axis.

Figure 4.24a shows the positive directions of the bending moments
M_z and M_y. These are really components of the resultant moment M.
Figure 4.24b shows the components and moment as vectors. The angle
α, given by

$$\tan \alpha = \frac{M_y}{M_z}$$

indicates the direction of the plane of the resultant moment M. Further,

$$\tan \alpha = \frac{M_y}{M_z}$$

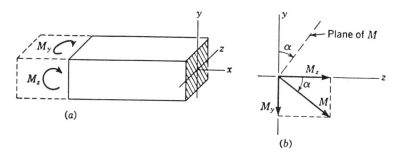

FIGURE 4.24 (a) Positive moments; (b) components of M.

The direction of the neutral axis can then be given by

$$\tan \theta = - \frac{M_y}{M_z} \frac{I_z}{I_y} = - \frac{I_z}{I_y} \tan \alpha$$

which shows that the neutral axis is not always perpendicular to the plane of the moment.

The surface that represents the stress distribution in Fig. 4.23a is a pair of planes. Therefore, the stress is proportional to the distance from the neutral axis and the maximum stress occurs at the greatest distance from the neutral axis.

4.16 TORSION SUPERPOSED ON FLEXURE

When a torsional stress pattern is superposed on a flexural stress pattern, the result is shear and normal stress at a typical point in a cross section of a bar. For example, Fig. 4.25 shows a cross section of a circular bar that is subjected to the couples M and T. At the point A, the shear stress is tangent to the boundary, and the magnitude is

$$\tau = \frac{Ta}{J}$$

Also, the flexural stress is compressive; its magnitude is

$$\sigma = - \frac{Ma}{I}$$

The state of stress at A is shown in Fig. 4.25b. The shear stress would be positive.

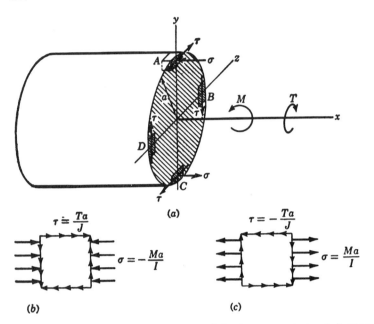

FIGURE 4.25 (a) Circular bar; (b) stresses at A; (c) stresses at B.

At point C in Fig. 4.25a, the flexural stress is tensile and is therefore considered to be positive. Figure 4.25c shows the state of stress at C, looking downward. The shear stress is negative.

At points B and D in Fig. 4.25a, the flexural stress is zero because these points are on the neutral axis of the bar. However, the shear stress is the algebraic sum of (1) the shear stress due to the transverse force,

$$\tau = \frac{VQ}{It}$$

and (2) the torsional shear stress

$$\tau = \frac{Ta}{J}$$

Principal stresses, maximum shear stress, and so on, can be calculated for any point by using the theories presented in the following sections.

4.17 STATE OF STRESS AT A POINT

Materials often fail on surfaces other than those on which failure might be anticipated from a simple viewpoint. For example, Fig. 4.26 shows an aluminum-alloy tension specimen after failure; here one might expect the failure to be on a surface perpendicular to the axis of the bar, but instead, the failure is on a surface at an angle of approximately 45° to the axis of the bar.

Figure 4.27a shows a rectangular element that is subjected to the stresses σ_x, σ_y, and τ_{xy}. Figure 4.27b shows a wedge cut from the rectangle in Fig. 4.27. The stress condition on the plane AA is described completely by the stresses σ_n and τ_{nt}.

A normal stress is positive if it is a tensile stress, and it is negative if it is a compressive stress. A shear stress, such as τ_{nt} in Fig. 4.27b, is positive if directed away from the vertex of the angle θ; it is negative if directed toward the vertex of the angle.

The four faces of the rectangle in Fig. 4.27a are designated by values of θ as follows, together with the algebraic signs of the stresses on the faces:

	Angle, θ	σ	τ
Top face	0°	+	+
Left-hand face	90°	+	−
Bottom face	180°	+	+
Right-hand face	270°	+	−

The normal stress σ_n can be found by writing the equation of equilibrium for the sum of the forces in the direction of σ_n in Fig.

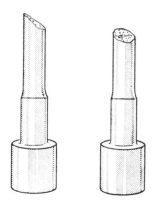

FIGURE 4.26

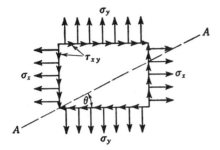

FIGURE 4.27a

4.27b. Thus, taking the dimension as unity in the direction perpendicular to the figure, and remembering that force is the product of stress and area, we have

$$\sigma_n \, ds(1) - \sigma_x \, dy \, (1) \sin \theta - \sigma_y \, dx \, (1) \cos \theta + \tau_{xy} \, dy \, (1) \cos \theta$$
$$+ \tau_{xy} \, dx \, (1) \sin \theta = 0$$

Dividing by ds and transposing yields

$$\sigma_n = \sigma_x \frac{dy}{ds} \sin \theta + \sigma_y \frac{dx}{ds} \cos \theta - \tau_{xy} \frac{dy}{ds} \cos \theta - \tau_{xy} \frac{dx}{ds} \sin \theta$$

and noting that dx/ds is cos θ and that dy/ds is sin θ,

$$\sigma_n = \sigma_x \sin^2 \theta + \sigma_y \cos^2 \theta - 2\tau_{xy} \sin \theta \cos \theta$$

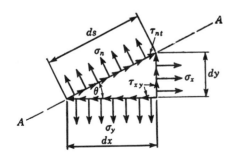

FIGURE 4.27b

or, in terms of the double angle,

$$\sigma_n = \frac{\sigma_x + \sigma_y}{2} + \frac{\sigma_y - \sigma_x}{2} \cos 2\theta - \tau_{xy} \sin 2\theta \tag{5}$$

The maximum and minimum values of σ_n, as θ is allowed to vary, can be found by equating to zero the derivative of σ_n with respect to θ. Thus

$$\frac{d\sigma_n}{d\theta} = -2 \frac{\sigma_y - \sigma_x}{2} \sin 2\theta - 2\tau_{xy} \cos 2\theta = 0$$

from which

$$\tan 2\theta = -\frac{\tau_{xy}}{(\sigma_y - \sigma_x)/2} \tag{6}$$

the directions in which σ_n is maximum and minimum. Equation (6) gives two values of 2θ which are 180° apart and, consequently, two values of which are 90° apart. When these angles are substituted in Eq. (5), the result is

$$\left. \begin{matrix} \text{Max.} \\ \text{Min.} \end{matrix} \right\} \sigma_n = \frac{\sigma_x + \sigma_y}{2} \pm \sqrt{\left(\frac{\sigma_y - \sigma_x}{2}\right)^2 + \tau_{xy}^2} = \left\{ \begin{matrix} \sigma_1 \\ \sigma_2 \end{matrix} \right. \tag{7}$$

Here the maximum value of σ_n is σ_1 and is obtained by using the positive sign in front of the radical; the minimum value of σ_n is σ_2 and is obtained by using the negative sign.

In calculating maximum and minimum normal stresses, it is customary to designate a tensile stress as positive and a compressive stress as negative, both in substituting for σ_y and σ_x and in interpreting the results.

The shear stress τ_{nt} shown in Fig. 4.27b can be found by writing the equilibrium equation for the sum of the forces in the direction of τ_{nt}. Thus

$$\tau_{nt} \, ds \, (1) + \sigma_x \, dy \, (1) \cos \theta - \sigma_y \, dx \, (1) \sin\theta + \tau_{xy} \, (1) \sin \theta$$

$$- \tau_{xy} \, dx \, (1) \cos \theta = 0$$

Dividing by ds and transposing gives us

$$\tau_{nt} = -\sigma_x \frac{dy}{ds} \cos \theta + \sigma_y \frac{dx}{ds} \sin \theta - \tau_{xy} \frac{dy}{ds} \sin \theta + \tau_{xy} \frac{dx}{ds} \cos \theta$$

Then, noting again that dy/ds is sin θ and dx/ds is cos θ, we have

$$\tau_{nt} = (\sigma_y - \sigma_x) \sin \theta \cos \theta + \tau_{xy}(\cos^2 \theta - \sin^2 \theta)$$

or, in terms of the double angle,

$$\tau_{nt} = \frac{\sigma_y - \sigma_x}{2} \sin 2\theta + \tau_{xy} \cos 2\theta \qquad (8)$$

The maximum and minimum values of τ_{nt} as θ is allowed to vary can be found by equating to zero the derivative of τ_{nt} with respect to θ. Thus

$$\frac{d\tau_{nt}}{d\theta} = 2\frac{\sigma_y - \sigma_x}{2} \cos 2\theta - 2\tau_{xy} \sin 2\theta = 0$$

from which

$$\tan 2\theta = \frac{(\sigma_y - \sigma_x)/2}{\tau_{xy}} \qquad (9)$$

the directions in which τ_{nt} is maximum and minimum. When these angles are substituted in the equation for τ_{nt}, the result is

$$\left.\begin{matrix}\text{Max.}\\\text{Min.}\end{matrix}\right\} \tau_{nt} = \pm \sqrt{\left(\frac{\sigma_y - \sigma_x}{2}\right)^2 + \tau_{xy}^2} \qquad (10)$$

Here the maximum value of τ_{nt} is obtained by using the positive sign and the minimum value by using the negative sign. However, the two stresses have the same absolute value; the only difference between them is a reversal of direction and is not considered important; hence it is customary to speak only of the maximum stress.

The maximum and minimum values of σ_n are called *principal stresses*. They occur on perpendicular planes and these planes are called *principal planes*. Further, it can be shown, by substituting the angles of Eq. (6) in Eq. (8) for the shear stress, that the shear stress is zero on a principal plane. Figure 4.28a shows the stresses on arbitrary planes through a point, and Fig. 4.28b shows the principal stresses at the point.

Since the right-hand side of Eq. (9) is the negative reciprocal of the right-hand side of Eq. (6), the angles 2θ given by one equation are 90° different from the angles 2θ given by the other equation. Therefore, the angles θ differ by 45°, and the planes of maximum

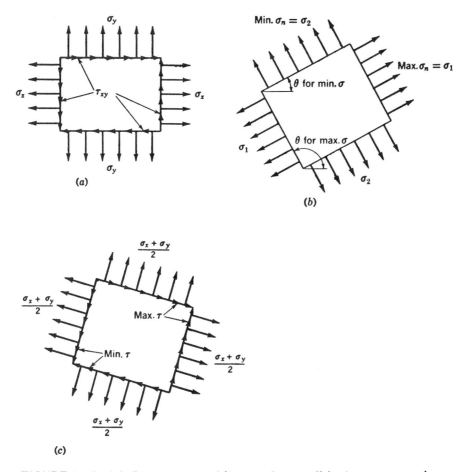

FIGURE 4.28 (a) Stresses on arbitrary planes; (b) stresses on principal planes; (c) stresses on planes of maximum and minimum shear stresses.

shear stress are at 45° with the principal planes. Furthermore, when the angles for the planes of maximum shear stress, from Eq. (9), are substituted in Eq. (5) for the normal stress σ_n, the result is

$$\bar{\sigma} = \frac{\sigma_x + \sigma_y}{2}$$

which shows that the normal stresses on the planes of maximum and minimum shear stress are each equal to the mean of σ_x and σ_y. Figure

4.28c shows the stress condition on the planes of maximum and minimum shear stress.

Ordinarily, one can determine by inspection which principal stress acts on which principal plane by drawing such pictures as Figs. 4.28a and 4.28b. Then the maximum shear stress must be on a plane at an angle of 45° with the principal planes.

However, it is possible to identify a certain principal stress with the proper principal plane by using the fact that the second derivative of a quantity is negative at a maximum point and positive at a minimum point. The second derivative of σ_n with respect to θ is

$$\frac{d^2 \sigma_n}{d\theta^2} = -2(\sigma_y - \sigma_x) \cos 2\theta + 4\tau_{xy} \sin 2\theta$$

Also,

$$\frac{d^2 \tau_{nt}}{d\theta^2} = -2(\sigma_y - \sigma_x) \sin 2\theta - 4\tau_{xy} \cos 2\theta$$

When values of 2θ have been established, it can be determined whether a given direction is to be associated with a maximum or minimum stress by seeing whether the second derivative is negative or positive, respectively.

The stresses acting on an element of material do work on the element as the element is deformed. The work is stored in the material as potential energy and represents work that the material can do as the stresses are removed. This type of potential energy is called *strain energy*. Work done by a force is the product of the force and the distance through which it acts.

The strain energy per unit volume of material can be calculated by using Hooke's law for the relation between stresses and strains. Fgiure 4.29 shows a rectangle that has the shape ABCD before being subjected to stress. The rectangle deforms to the shape $AB_1C_1D_1$ when the stresses σ_x and σ_y are applied. The rectangle $AB_2C_2D_2$ are intermediate configurations and show small increments of the deformation. The work done by σ_x, at the right-hand face of the element, during the deformation $d\varepsilon_x \, dx$ is

$$\sigma_x \, dy(1) \, d\varepsilon_x \, dx$$

and the work done by σ_y, on the top face of the element, is

$$\sigma_y \, dx \, (1) \, d\varepsilon_y \, dy$$

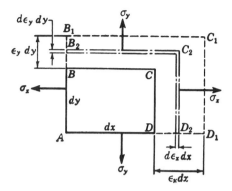

FIGURE 4.29

The strain energy per unit volume in this small deformation is

$$dU_1 = \frac{\sigma_x \, dy \, d\varepsilon_x \, dx + \sigma_y \, dx \, d\varepsilon_y \, dy}{dx \, dy \, (1)} = \sigma_x \, d\varepsilon_x + \sigma_y \, d\varepsilon_y$$

The differentials of the strain components are, from Hooke's law,

$$d\varepsilon_x = d\left[\frac{1}{E}(\sigma_x - \nu\sigma_y)\right] = \frac{1}{E}(d\sigma_x - \nu \, d\sigma_y)$$

$$d\varepsilon_y = d\left[\frac{1}{E}(\sigma_y - \nu\sigma_x)\right] = \frac{1}{E}(d\sigma_y - \nu \, d\sigma_x)$$

Thus

$$dU_1 = \frac{1}{E}(\sigma_x \, d\sigma_x - \nu\sigma_x \, d\sigma_y) + \frac{1}{E}(\sigma_y \, d\sigma_y - \nu\sigma_y \, d\sigma_x)$$

The strain energy per unit volume, during the total deformation shown in Fig. 4.29, is

$$U_1 = \int dU_1 = \frac{1}{E}\int_0^{\sigma_x} \sigma_x \, d\sigma_x - \frac{\nu}{E}\int_0^{\sigma_y} \sigma_x \, d\sigma_y + \frac{1}{E}\int_0^{\sigma_y} \sigma_y \, d\sigma_y$$

$$- \frac{\nu}{E}\int_0^{\sigma_x} \sigma_y \, d\sigma_x$$

If it is assumed that σ_x and σ_y remain in the same ratio to each other, the second and fourth integrals can be written so that

$$U_1 = \frac{1}{E} \int_0^{\sigma_x} \sigma_x \, d\sigma_x - \frac{\nu}{E} \frac{\sigma_y}{\sigma_x} \int_0^{\sigma_y} \sigma_y \, d\sigma_y + \frac{1}{E} \int_0^{\sigma_y} \sigma_y \, d\sigma_y$$

$$- \frac{\nu}{E} \frac{\sigma_y}{\sigma_x} \int_0^{\sigma_x} \sigma_x \, d\sigma_x$$

$$= \frac{1}{E} \frac{\sigma_x^2}{2} - \frac{\nu}{E} \frac{\sigma_x}{\sigma_y} \frac{\sigma_y^2}{2} + \frac{1}{E} \frac{\sigma_y^2}{2} - \frac{\nu}{E} \frac{\sigma_y}{\sigma_x} \frac{\sigma_x^2}{2}$$

$$= \frac{1}{E} \frac{\sigma_x^2}{2} + \frac{\sigma_y^2}{2} - \nu \sigma_x \sigma_y \qquad (11)$$

The strain energy due to shear stresses can be calculated by using Fig. 4.30, which shows a rectangle in the shape ABCD before being subjected to shear stresses τ_{xy}. The final configuration is $AB_1C_1D_1$; $AB_2C_2D_2$ represents two intermediate configurations.

The shear strain is

$$\gamma = \alpha_1 + \alpha_2$$

and the increment of shear strain in the configurations $AB_2C_2D_2$ is

$$d\gamma = d\alpha_1 + d\alpha_2$$

The work done by the force on the right-hand face of the element, in the increment of deformation, is

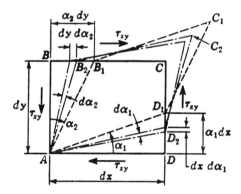

FIGURE 4.30

$$\tau_{xy} \; dy \; (1) \; dx \; d\alpha_1$$

and the work done by the force on the top of the element is

$$\tau_{xy} \; dx \; (1) \; dy \; d\alpha_2$$

Thus the strain energy per unit volume, in the increment of deformation, is

$$dU_2 = \frac{\tau_{xy} dy \; dx \; d\alpha_1 + \tau_{xy} \; dx \; dy \; d\alpha_2}{dx \; dy}$$

$$= \tau_{xy}(d\alpha_1 + d\alpha_2) = \tau_{xy} \; d\gamma$$

The quantity $d\gamma$ is

$$d\gamma = d \; \frac{\tau_{xy}}{G} = \frac{d\tau_{xy}}{G}$$

so that

$$dU_2 = \tau_{xy} \; \frac{d\tau_{xy}}{G}$$

Therefore, the strain energy of shear per unit volume, for the total deformation, is

$$U_2 = \int dU_2 = \frac{1}{G} \int_0^{\tau_{xy}} \tau_{xy} \; d\tau_{xy} = \frac{\hat{\tau}_{xy}^2}{2G} \tag{12}$$

When both normal stresses and shear stresses occur, the total strain energy per unit volume is

$$U = U_1 + U_2 = \frac{1}{E} \left(\frac{\sigma_x^2}{2} + \frac{\sigma_y^2}{2} - \nu \sigma_x \sigma_y \right) + \frac{1}{G} \frac{\tau_{xy}^2}{2} \tag{13}$$

Since stress and modulus of elasticity are expressed in the same units, strain energy per unit volume is in the same units as stress, usually pounds per square inch (psi). This can be interpreted as inch-pounds per cubic inch.

4.18 STATE OF STRAIN AT A POINT

A theory of the state of strain at a point is useful in analyzing strain
measurements and in seeking to understand the failure of materials.
Given the strains in certain directions, the theory can be used to calculate the strain in any other chosen direction. Figure 4.31a shows an
element ABCD before straining and $AB_1C_1D_1$ after straining. The nor-
mal strains are ε_x and ε_y, and the shear strain is $\gamma_{xy} = \alpha_1 + \alpha_2$. Figure 4.31b shows an element that is oriented at an angle θ with the x
axis; the configuration of the element is ABCD before straining and
$AB_1C_1D_1$ after staining. Here the normal strains are ε_n and ε_t and
the shear is $\gamma_{nt} = \alpha_3 + \alpha_4$.

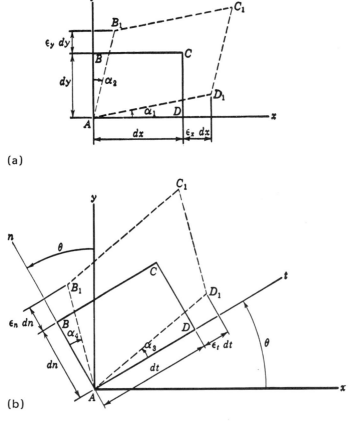

(a)

(b)

FIGURE 4.31

The normal strain in a direction given by θ can be found by juggling equations already presented. Thus the normal stress in the direction n in Fig. 4.31b is

$$\sigma_n = \frac{\sigma_x + \sigma_y}{2} + \frac{\sigma_y - \sigma_x}{2} \cos 2\theta - \tau_{xy} \sin 2\theta$$

Then the normal stress in the direction t in Fig. 4.31b is obtained by using the same expression, with θ replaced by $\theta - 90°$.

$$\sigma_t = \frac{\sigma_x + \sigma_y}{2} + \frac{\sigma_y - \sigma_x}{2} \cos 2\theta + \tau_{xy} \sin 2\theta$$

Hooke's law can be expressed as

$$\varepsilon_n = \frac{1}{E} (\sigma_n - \nu\sigma_t)$$

which gives

$$\varepsilon_n = \frac{1}{E} \left[(1 - \nu) \frac{\sigma_x + \sigma_y}{2} + (1 - \nu) \frac{\sigma_y - \sigma_x}{2} \cos 2\theta - (1 + \nu)\tau_{xy} \sin 2\theta \right]$$

Next, Eq. (2) can be rewritten as

$$\sigma_x = \frac{E}{1 - \nu^2} (\varepsilon_x + \nu\varepsilon_y)$$

$$\sigma_y = \frac{E}{1 - \nu^2} (\varepsilon_y + \nu\varepsilon_x)$$

(14)

Use of this pair of equations and Eq. (1) leads to

$$\varepsilon_n = \frac{\varepsilon_x + \varepsilon_y}{2} + \frac{\varepsilon_y - \varepsilon_x}{2} \cos 2\theta - \frac{(1 + \nu)G}{E} \gamma_{xy} \sin 2\theta$$

Finally, using Eq. (3) for the relation between E and G, we have

$$\varepsilon_n = \frac{\varepsilon_x + \varepsilon_y}{2} + \frac{\varepsilon_y - \varepsilon_x}{2} \cos 2\theta - \frac{\gamma_{xy}}{2} \sin 2\theta \qquad (15)$$

The shear stress for the element in Fig. 4.31b is τ_{nt}. From Eq. (8), this is

$$\tau_{nt} = \frac{\sigma_y - \sigma_x}{2} \sin 2\theta + \tau_{xy} \cos 2\theta$$

From Hooke's law,

$$\gamma_{nt} = \frac{\tau_{nt}}{G} = \frac{\sigma_y - \sigma_x}{2G} \sin 2\theta + \frac{\tau_{xy}}{G} \cos 2\theta$$

Then, using Eqs. (14) and (2) again, we have

$$\gamma_{nt} = \frac{E}{(1 + \nu)G} \frac{\varepsilon_y - \varepsilon_x}{2} \sin 2\theta + \gamma_{xy} \cos 2\theta$$

4.19 THE FINITE ELEMENT METHOD

The finite element method is a numerical analysis technique for obtaining approximate solutions to a wide range of scientific, engineering, and design problems. The method was originally developed to study stresses in complex airframe strucutres, but it has now been applied to the broad field of continuum mechanics.

Today, it is becoming more important to obtain approximate numerical solutions to engineering and design problems rather than exact solutions from formulas and equations. For example, one would like to know the load capacity of a plate that has various size holes and stiffeners or the rate of liquid flow through an opening of an arbitrary shape. To solve such type of problems, one would proceed to write down the governing equations and their boundary conditions, realizing immediately that no simple analytical solution is available. The difficulty lies in the fact that either the geometry or some other condition of the problem is variable or arbitrary. Analytical solutions to these problems seldom exist, although these are the type of problems that analysts (scientists, engineers, and designers) are called on to solve.

To solve such problems, the analyst could simplify the problem by ignoring the difficulties and reducing the problem to one that can be handled. This procedure sometimes works but usually leads to serious inaccuracies or worse, wrong answers. Today, with the advent of large-scale digital computers, it is more logical to retain the complexities of the problem and attempt to find an approximate numerical solution to the problem.

There are several approximate numerical analysis methods in use today, the most commonly being the general finite different method.

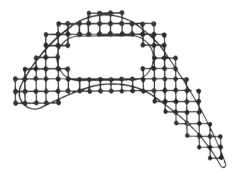 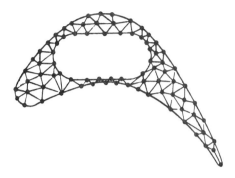

FIGURE 4.32a   FIGURE 4.32b

The finite difference model of a problem gives a pointwise approximation to the governing equations. This model, which is formed by writing difference equations for an array of grid points, is improved as
more points are used. When irregular geometries or unusual boundary
conditions exist, it is found that finite difference techniques become
difficult to use.

A numerical method known as the *finite element method* has been
devised which divides the model into many small, interconnected sub-
regions known as *elements*. A finite element model of a problem gives
a piecewise approximation to the governing equations. The basic approach of the finite element method is that a solution region can be analytically modeled or approximated by replacing it with an assemblage
of discrete elements. These elements can be used to represent exceed-
ingly complex geometric shapes, since these elements can be put together in various ways.

Figures 4.32a and 4.32b show cross sections of a turbine blade and
how a finite difference model (Fig. 4.32a) and a finite element model
(Fig. 4.32b) would be used to represent them. A uniform finite difference mesh would cover the blade adequately, but the boundaries
must be approximated by a series of horizontal and vertical lines known
as stair steps. The boundaries are better approximated by using the
simplest two-dimensional element, the triangle, which can be done when
using the finite element model.

In a continuum (a region of space in which a particular phenomenon
is occurring) problem of any dimension, the field variable (pressure,
temperature, displacement, stress, or some other quantity) possesses
infinitely many values because it is a function of each generic point in
the body or solution region. Consequently, the problem is one with an
infinite number of unknowns. The finite element discretization procedures reduce the problem to one of a finite number of unknowns by

dividing the solution region into elements used and by expressing the unknown field variable in terms of assumed approximating functions within each element. The approximating functions (sometimes called interpolation functions) are defined in terms of the vlaues of the field variables at specified points called *nodes* or *nodal points*. Nodes usually lie on the element boundaries where adjacent elements are considered to be connected. In addition to boundary nodes, an element may also have a few interior nodes. The nodal values of the field variable and the interpolation functions for the elements completely define the behavior of the field variable within the elements. For the finite element representation of a problem, the nodal values of the field variable become the new unknowns. Once these unknowns are found, the interpolation functions define the field variable throughout the assemblage of elements.

Clearly, the nature of the solution and the degree of approximation depend not only on the size and number of the element used, but also on the interpolation functions selected. As one would expect, these functions cannot be chosen arbitrary because certain compatibility conditions should be satisfied. Often functions are chosen so that the field variable or its derivatives are continuous across adjoining element boundaries.

The most important feature of the finite element method, which sets it apart from other approximate numerical methods, is the ability to formulate solutions for individual elements before putting them together to represent the entire problem. This means, for example, that if we are treating a problem in stress analysis, the force-displacement or stiffness characteristics of each individual element can be found and then these elements can be assembled to find the stiffness of the whole structure. In essence, a complex problem reduces to considering a series of greatly simplified problems.

Another advantage of the finite element method is the variety of ways in which one can formulate the properties of individual elements. There are basically four different approaches. The first approach to obtaining element properties is called the *direct approach* becuase its origin is traceable to the direct stiffness method of structural analysis. Although the direct approach can be used only for relatively simple problems, it is the easiest to understand when meeting the finite element method for the first time. The direct approach also suggests the need for matrix algebra in dealing with the finite element equations.

Element properties obtained by the direct approach can also be determined by the more versatile and more advanced *variational approach*. The variational approach relies on the calculus of variations and involves extremizing a functional. For problems in solid mechanics, the functional turns out to be the potential energy, the complementary potential energy, or some derivative of these, such as the

Reisser variational principle. Knowledge of the variational approach is necessary to work beyond the introductory level and to extend the finite element method to a wide variety of engineering and design problems. Whereas the direct approach can be used to formulate element properties for only the simplest element shapes, the variational approach can be employed for both simple and sophisticated element shapes.

A third and even more versatile approach to deriving element properties has its basis entirety in mathematics and is known as the *weighted residuals approach*. The weighted residuals approach begins with the governing equations of the problem and proceeds without relying on a functional or a variational statement. This approach is advantageous because it thereby becomes possible to extend the finite element method to problems where no functional is available. For some problems, there is no functional, either because one may not have been discovered or because one does not exist.

A fourth approach relies on the balance of thermal and/or mechanical energy of a system. The *energy balance approach* requires no variational statement and thus broadens considerably the range of possible applications of the finite element method.

Regardless of the approach used to find the element properties, the solution of a continuum problem by the finite element method always follows an orderly step-by-step process. To summarize in general terms how the finite element method works, the following steps are listed.

1. *Discretize the continuum.* The first step is to divide the continuum or solution region into elements. In the example of Fig. 4.32b the turbine blade has been divided into triangular elements which might be used to find the stress distribution in the blade. A variety of element shapes may be used, and, with care, different element shapes may be employed in the same solution region. Indeed, when analyzing an elastic structure that has different types of components such as plates and beams, it is not only desirable but also necessary to use different types of elements in the same solution. Although the number and the type of elements to be used in a given problem are matters of engineering and design judgment, the analyst can rely on the experience of others for guidelines.

2. *Select interpolation functions.* The next step is to assign nodes to each element and then choose the type of interpolation function to represent the variation of the field variable over the element. The field variable may be a scalar, a vector, or a higher-order tensor. Often, although not always, polynomials are selected as interpolation functions for the field variable because they are easy to integrate and differentiate. The degree of the polynomial chosen depends on the

number of nodes assigned to the element, the nature and number of unknowns at each node, and certain continuity requirements imposed at
the nodes and along the element boundaries. The magnitude of the
field variable as well as the magnitude of its derivatives may be the unknowns at the nodes.

   3. *Find the element properties.* Once the finite element model has
been established (i.e., once the elements and their interpolation functions have been selected), the matrix equations can be determined, thus
expressing the properties of the individual elements. For this task,
use is made of one of the four approaches just mentioned: the direct
approach, the variational approach, the weighted residual approach, or
the energy balance approach. The variational approach is often the
most convenient, but for any application the approach used depends
entirely on the nature of the problem.

   4. *Assemble the element properties to obtain the system equations.*
To find the properties of the overall system modeled by the network of
elements, one must assemble all the element properties. In other words,
one must combine the matrix equations expressing the behavior of the
elements and form the matrix equations expressing the behavior of the
entire solution region or system. The matrix equations for the system
have the same form as the equations for an individual element except
that they contain many more terms because they include all nodes.

   The basis for the assembly procedure stems from the fact that at a
node where elements are interconnected, the value of the field variable
is the same for each element sharing that node. Assembly of the element equations is a routine matter in finite element analysis and is usually done by a digital computer. Before the system equations are
ready for solution, they msut be modified to account for the boundary
conditions of the problem.

   5. *Solve the system equations.* The assembly process of the preceding step gives a set of simultaneous equations that must be solved
to obtain the unknown nodal values of the field variable. If the equations are linear, a number of standard solution techniques can be used.

   6. *Make additional computations if desired.* Sometimes it is desirable to use the solution of the system equations to calculate other
important parameters. For example, in a stress problem such as the
plate problem, the solution of the system equations gives shear and
normal stress distribution within the system. From the nodal values
of the stress, strain distributions can be calculated or perhaps maximum shear or normal stresses if these are desired.

   Finally, from a practioners' viewpoint, the finite element method,
like any other numerical analysis technique, can always be made more
efficient and easier to use. As the method is applied to larger and more
complex problems, it becomes increasingly important that the solution
process remain economical. This means that studies to find better ways
to solve simultaneous linear and nonlinear equations will certainly continue. Also, since implementation of the finite element method usually

requires a considerable amount of data handling, it is expected that ways to automate this process and make it more error-free will evolve. Finite element modeling using interactive computer graphics is such a process.

Interactive computer graphics finite element modeling (FEM) is a preprocessing step for finite element analysis (FEA) and a postprocessing display of the results. Here are the six simple steps involved when using FEM:

1. *Test shape is overlaid with a mesh of points.* These points may be assigned automatically as a mesh generated by the system, or semiautomatically with user intervention in areas of special interest (stress or temperature concentration, for example), or entirely user controlled in cases of extreme complexity. Meshes may be generated for essentially any engineering/design shape under consideration, including lines, arcs, splines, conics, ruled surfaces, surfaces of revolution, Coon's patches, B-surfaces, and fillet surfaces.

2. *Verification is interactive visual.* In addition to the screen, the user can command immediate hardcopy for discussion or "desk" review later. Manually introduced changes, visually verified, reflect detailed design or attribute criteria.

3. *Nongraphical attribute data can now be added to each individual point on the mesh, collectively or individually:* For example, pressure over an entire area, or temperature as a static value or gradient. Such attribute data, as with the original mesh generation, may be assigned automatically to all points, semiautomatically (with user intervention) or entirely by user control. All attribute data are stored. Complete printout listings permit verification from the screen or from the hard copy.

4. *Collected data, verified by the user, can be processed by the FEM module for FEA analysis in any of the popular FEA program codes, as commanded by the user.* These codes include ANSYS, MSC/NASTRAN, SUPERB, and so on—virtually every current, widely used FEA analysis concept. The FEM files are now ready for FEA processing on the user's mainframe or time-shearing service. The appropriate job control language interface may be added, based on mainframe type (accounting data, password, job number, etc.).

5. *FEA results from the large system can now be displayed on the interactive computer graphics system by the FEM software.* Such displays can include "before" and "after" representations of stress effects on specific parts, in different magnifications, under user command. Now the scientist, engineer, or designer can see exactly what the part will do when in use.

6. *Changes and revisions can be accomplished rapidly,* by revising existing data, stored in the system, automatically or manually. The graphic and hard-copy display of FEA results provides immediate, powerful insights to designers.

4.20 FINITE ELEMENT MODELING USING
INTERACTIVE COMPUTER GRAPHICS

The geometry used in defining the finite element model (FEM) is obtained directly from the three-dimensional graphics model (Chapter 2) created during the conceptual design stage. In other words, when a person constructs a part on an interactive computer graphics system (ICGS), they create its associate data base. This data base of the part is used to develop and define its finite element model on an ICGS. This geometric model is next divided into many discrete sections called *finite elements*; the combination of these elements is known as a *finite element mesh*.

By using interactive computer graphics, two- or three-dimensional finite element meshes are generated over the most complex model geometry. The nodes can be uniformly distributed over the model, or they may be spaced more closely in areas where stresses would be critical or higher accuracy is required. The process of inserting nodes (grid points) is interactive and fast. Faulty meshes are prevented from being created since the computer checks for ill-formed elements, duplicate nodes, and unused nodes. To perform the analysis, additional information is required. The engineer or designer interactively assigns a range of attributes, including loads, restraints, and material properties to any individual or group of nodes and elements.

This FEM is ready for the next step in the analysis process by claculating the behavior of the FEM due to the imposed loads. There are various analysis software available, such as NASTRAN, ANSYS, and so on. Usually, the FEM data base must be translated or fomatted so that it is compatible with the particular analysis data base format.

Model deformation caused by loads can be displayed as a deflected (defomred) shape, alone or superimposed on the undeflected model. Stresses and strain may also be presented graphically as color contour plots. These display features permit the engineer or designer to spot quickly areas of high stress and therefore areas of potential structural failure. Printed output is also available.

After analysis is performed, the results are returned to the data base. With this information readily available, the engineer or designer can more precisely evaluate alternative design concepts and select one or more for refinement or additional analysis. To ensure their ability to meet performance requirements, parts must undergo meticulous testing. Finite element model (FEM) and analysis (FEA) help designers and engineers to predict part behavior under varying conditions and level stress.

Because the system's integrated data base stores current information from all phases of design, the design process can move directly from geometric modeling to finite element modeling without time-consuming reconstruction of geometry or transcription error. Working di-

directly with the three-dimensional geometric model on a display screen, designers and engineers can make design changes instantly and generate a finite element mesh interactively or automatically.

Designers and engineers can ensure accuracy by using display capabilities to rotate and magnify portions of the mesh and highlight them with color. Specialized FEM software will shrink elements to verify connectivity. Once a satisfactory mesh has been created, material properties, constraints, pressures, temperatures, forces, and moments are defined for the nodes and elements in preparation for analysis.

Interactive computer graphic systems offer several ways to perform finite element analysis. Information from a completed finite element model is formatted by the system for input to the appropriate analysis program. Data can then be communicated to a mainframe computer for analysis, or analyzed on-line by using the system's own processor.

Finite element modeling and finite element analysis programs convert numerical data and analysis results to graphic form, enabling designers and engineers to examine and interpret them quickly. Engineers and designers can see two- or three-dimensional color-keyed representations of stress distributions or conditions. Dynamic simulat9ons can show the part deformation that results from transient loads, or the effects of fundamental modes of vibration. By providing a fast, efficient way to test alternatives and select the most efficient design, interactive computer graphics frees designers and engineers to maximize their creative skills and design/engineering expertise.

4.21 FINITE ELEMENT MODELING COMMANDS

The intent of the following sections of this chapter is to acquaint the reader with the procedure required to construct the finite element model using the graphics model previously created on an interactive computer graphics system. The Computervision Interactive Computer Graphics System with its CADDS FEM (Computer-Aided Drafting and Design Finite Element Modeling) software will be used to illustrate how such a system operates. The actual command syntax necessary to create the FEM with appropriate explanations is presented in the following sections. The reader must remember that the procedure for creating the finite element model on most interactive computer graphics systems is similar but that the actual syntax used on each system varies from one manufacturer to the other.

The following types of data (information) can be created with CADDS FEM:

1. Nodes (referred to as grid points in CADDS FEM)
2. Elements
3. Element property descriptions

4. Material descriptions
5. Loads
6. Constraints
7. Finite element analysis (FEA) compatibility with finite element model (FEM) data

4.22 NODES (GRID POINTS)

Nodes (grid points) are usually created first when the user begins construction of the finite element model on an ICGS since they are needed for defining the elements themselves. The operator can insert grid points at any location in model space and on the following CADDS entities:

1. Lines
2. Ruled surfaces
3. Arcs
4. Surfaces of revolution
5. B-spline surfaces
6. Composite curves
7. B-spline
8. Filleted surfaces
9. Strings
10. Tabulated cylinders
11. Nodel lines
12. Conics
13. Cpoles
14. Spoles

4.22.1 Grid Point Parameters

Grid point parameters consist of (1) the starting grid point number and (2) the increment by which succeeding grid point numbers are assigned. Two commands may be used to control the values assigned to the parameters: (1) INITIALIZE FEM and (2) SELECT FEM. Additionally, the SAVE FEM command sotres the values assigned to grid point parameters. The system performs this function automatically when the user files a FEM part.

The INITIALIZE FEM command prepares the CADDS data base for FEM use and must be issued at the start of every FEM session. If during a FEM session the user is creating a new part, INITIALIZE FEM assigns the associated values to the parameters. During subsequent sessions in which the user is working on the FEM part, INITIALIZE FEM retrieves the values that existed when the operator filed the part.

Example 1

#n# INITIALIZE FEM MASTER DATA.FEM.SPECIAL.EDF [CR]

This command initializes FEM parameters and specifies that DATA.FEM.
SPECIAL.EDF will be the master element description file from which
element types will be selected.

The SELECT FEM command allows the user to define new values
for FEM parameters. The new values become the defaults associated
with the FEM part. If the operator uses this command at any time
when working on a FEM part, the user need not reissue the command
every time he or she activates the part, because the INITIALIZES
FEM command will retrieve the most current values for each parameter,
including any values specified by the SELECT FEM command.

Example 2

#n# SELECT FEM GPNEXT 10 ELNEXT 10 [CR]

Assuming that the FEM part is new (i.e., no grid points or elements
have been inserted), this command line results in the following values
for FEM parts:

Parameter	Value
Next grid point number	10
Grid point number increment	1
Next element number	10
Element number increment	1
Master element description file	DATA.FEM.DEFAULT,EDF

The SAVE FEM command stores the most current values for FEM
parameters in the CADDS part, so the parameters will be up to date
when the user next works on the FEM part. The system issues this
command automatically whenever the user files a CADDS FEM part, so
there are few occasions when the user would actually have to use this
command. If the user does use the SAVE FEM command during a FEM
session, the user must use INITIALIZE FEM afterwards in order to retrieve FEM part parameters.

Example 3

#n# SAVE FEM [CR]

4.22.2 Creating Grid Points

The user may create grid points on a wire frame model, a model defined by surfaces, a combination wire frame/surface model, or at any

G I I
✱

FIGURE 4.33 Grid point and entity tag.

location in model space. As with all CADDS entity creation commands,
the user may insert grid points by digitizing or by entering coordinates. A grid point appears on the screen labeled with a tag. For example, grid point 11 will be tagged G11 (Fig. 4.33).

Grid points are not affected when the user rotates or zooms the
model; the tags are displayed parallel to the screen and thus are always visible unless the user choose not to display them. Three commands enable the user to create grid points: (1) INSERT GRID POINT,
(2) GENERATE GRID POINT, and (3) GENERATE MESH.

The INSERT GRID POINT command creates one or more grid points
at specified locations in model space, or on graphic entities (lines,
arcs, B-splines, surfaces, etc.). The user may also specify omitted
degrees of freedom, permanent single-point constraints, and grid
point rotation angles.

Example 4

a. #n# INSERT GPOINT ON:MODEL ent d_1 MODEL d_2 $\cdots$ d_n [CR]

Digitize d_1 identifies the entity on which grid points are to be inserted. Digitizes $d_2 \cdots d_n$ specify the positions nearest to which the
grid points are to be inserted. A normal dropped from the digitized
location onto the selected entity defines the positions of the grid
points.

b. #n# INSERT GPOINT PROJECT:MODEL ent d_1 MODEL end d_2d_3
    MODEL loc $d_4 \cdots d_n$ [CR]

Digitize d_1 identifies a surface, d_2 and d_3 specify a projection vector,
and $d_4 \cdots d_n$ identify the locations for grid points to be projected onto
the surface.

c. #n# INSERT GPOINT NUMBER 5-10 15 20-25:MODEL loc $d_1 \cdots d_{13}$
    [CR]

This command line creates 13 grid points. Note that the punctuation
for list/range format is similar to that used to identify existing grid
points; that is, ranges are specified by hyphens and numbers in a list
are separated by spaces. Assuming a grid point increment of 1 for a
new CADDS part, the next available grid point number is 26 (i.e., the
system will not use numbers 1−4, 11−14, or 16−19 unless the user
specifies them explicitly in another command).

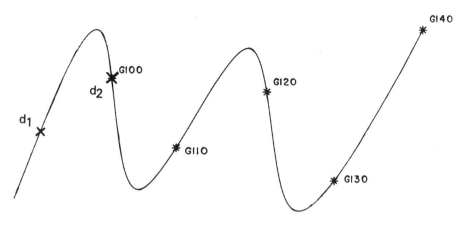

FIGURE 4.34

The GENERATE GPOINT command creates one or more grid points along a curve or composite curve. A composite curve can be made up pf different types of entities (consisting of up to 50 lines, arcs, conics, B-splines, strings, and nodal lines joined end to end).

Example 5

a. #n# GENERATE GPOINT ON N 5 NUMBER 100 INCREMENT 10 OMIT 1 STPT: MODEL ent d_1; DIGITIZE START POINT MODEL loc d_2 [CR] (see Fig. 4.34)
b. #n# GENERATE GPOINT ON N 5 ARITHMETIC 1 NUMBER 2 : MODEL ent d_1 [CR] (see Fig. 4.35)

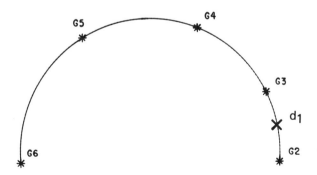

FIGURE 4.35

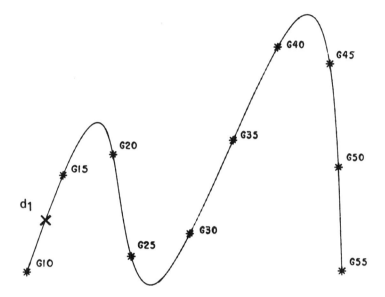

FIGURE 4.36

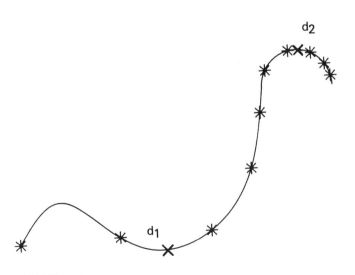

FIGURE 4.37

c. #n# GENERATE GPOINT ON N 10 NUMBER 10 PSPC 1 INCREMENT 5 : Model ent d_1 [CR] (see Fig. 4.36)
d. #n# GENERATE GPOINT ON N 10 DISTANCE 3.5 GEOMETRIC 0.75: MODEL ent d_1d_2 [CR] (see Fig. 4.37)

The GENERATE MESH command generates meshes automatically, or copies existing meshes by translation, rotation, or mirroring.

4.22.3 Editing and Deleting Grid Points

For editing and deleting operations, the user may identify existing grid points by digitizing them directly (using the grid point entity mask if appropriate) or the user may specify them by their tags in the Getdata portion of the command. Commands related to editing and deleting grid points are (1) CHANGE GRID POINT, (2) RENUMBER GRID POINT, (3) MERGE GRID POINT, and (4) DELETE ENTITY.

The CHANGE GRID POINT command changes old values or adds new values to attributes attached to grid points (i.e., omitted degrees of freedom and permanent single-point constraints). The user may also use this command to renumber grid points and to convert cartesian grid points to cylindrical or spherical grid points. These grid point types are not different in appearance from cartesian grid points.

Example 6

a. #n# CHANGE GPOINT START 80 INCREMENT 2 : MODEL ent $d_1 \cdots d_n$ [CR]

Digitizes $d_1 \cdots d_n$ identify the grid points, which are renumbered to 80, 82, 84, 86, and so on.

b. #n# CHANGE GPOINT SPHERICAL : MODEL ent TAG G40-45 G50-55 [CR]

This command line converts 12 grid points to spherical grid points.

The RENUMBER GRID POINT command renumbers grid points. The original grid points numbers are changed by a constant increment or are replaced by a new sequence of numbers which the user specifies in the command.

Example 7

a. #n# RENUMBER GPOINT TO 95-100 : MODEL ent TAG G45-50 [CR]
   This command renumbers grid points 45−50 to 95−100.
b. #n# RENUMBER GPOINT INCREMENT 10 : MODEL ent TAG G10-12 [CR]
   This command renumbers grid point 10−12 to 20−22.

The MERGE GRID POINT command checks and merges coincident grid points. It can also be used for checking and reporting the coincident grid points with no merge operation. Once the coincident grid points are merged, all FEM relationships are updated to maintain the integrity of the FEM data.

Example 8

a. #n# MERGE GPOINT : MODEL ent $d_1 \cdots d_n$ [CR]
    This command checks (and possibly merges) 10 digitized grid points.
b. #n# MERGE GPOINT SEQA 1-10 [CR]
    Here grid points specified by the tags 1−10 are checked (and possibly merged).
c. #n# MERGE GPOINT SEQA 20 22-30 : MODEL ent $d_1 \cdots d_{13}$ [CR]

The DELETE ENTITY command deletes grid points. It removes all information about the specified grid points from the part and updates all references and associations to maintain the integrity of the FEM part.

4.22.4 Controlling Grid Point Appearance

Due to the complexity of most finite element models, it is often useful to change the appearance of FEM entities for greater visual clarity:

1. The user may display or remove grid point tags from the screen using the SELECT TAG and ECHO TAG commands.
2. The user may also blank out the grid points themselves using the BLANK ENTITY command. To redisplay blanked grid points, use UNBLANK ENTITY.
3. Additionally, the user may place grid points on a layer other than that used for elements. To do this, use the SELECT LAYER or CHANGE LAYER commands. Ther user may then echo the grid points or remove them from the screen using the ECHO LAYER command.

4.22.5 Grid Point Status Information

Three commands provide the user with information about grid points in the FEM part:

1. LIST FEM provides both summary and detail information on grid points.
2. MERGE GRID POINT can be used to list the coincidence of grid points without merging them.
3. HIGHLIGHT FEM creates displays in which grid points with particular characteristics are highlighted.

4.23 ELEMENTS

Elements represent the basic structure of the finite element model.
In CADDS FEM, elements do not contain coordinate information but reference the coordinates of grid points instead. The user can choose
from several methods to create elements: (1) use existing grid points
and interactively insert elements, (2) use existing grid points and
generate elements automatically, (3) use existing elements and grid
points and copy them, and (4) define boundaries within which grid
points and elements are created automatically.

4.23.1 Element Parameters

Element parameters consist of the starting element number and the
increment by which succeeding element numbers are assigned. Two
commands may be used to control the values assigned to the two parameters, which are INITIALIZE FEM and SELECT FEM. The INITIAL-
IZE FEM command sets up default vlaues, while the SELECT FEM command allows the user to speicfy other values for element parameters.
Additionally, the SAVE FEM command stores the values assigned to
FEM element parameters. The system performs this function automatically whenever the user files a FEM part. CADDS FEM provides
a library of 54 element type and also allows the user to create new element types by using the DEFINE FEMSYSTEM command.

4.23.2 Element Creation

The user may insert elements by digitizing grid points or by referring
to grid point tag numbers. The system displays an element together
with an element tag, which is similar to a grid point tag. For example,
element number 29 will be displayed with the tag E29 (Fig. 4.38 shows
a three-node beam). Note that the tag number of an element appears
at the center of the element. In other respects the display characteristics of element tags are exactly like those of grid point tags. The
following is a summary of the element creation commands: (1) INSERT
ELEMENT, (2) GENERATE ELEMENT, and (3) GENERATE MESH.

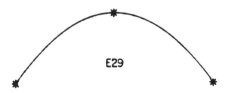

FIGURE 4.38 Element and entity tag.

The INSERT ELEMENT command requires that the user specify the type of element to be inserted. The user must also digitize (or reference by tag) the grid points that define the position of the element. The order in which the user digitize the grid points is important for the proper representation of the element in the CADDS data base and in the analysis program the user is using (see Figs. 4.39 through 4.41).

The first time the user inserts a type of element which the user has not used previously in the model, the system retrieves the element from the master element description file and adds it to the local element file for the part. For example, suppose that the user inserts a tetrahedron for the first time; the system would display the message TETR4 WAS ADDED TO LOCAL FILE. The second time the user inserts a tetrahedron, the system will retrieve it from the local file (which is associated with the user's FEM part) and will not display this message.

Example 1

#n# INSERT ELEMENT TYPE NSTIF503 [CR]
[CR]
: MODEL ent 4 NODES TAG G10-13 [CR]   (see Fig. 4.42)

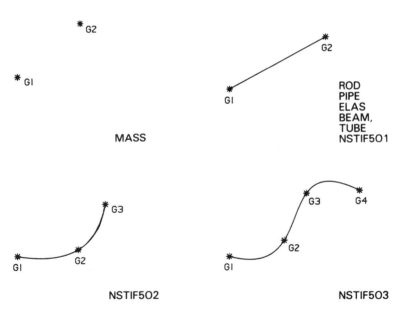

FIGURE 4.39 CADDS 4 one-dimensional finite elements.

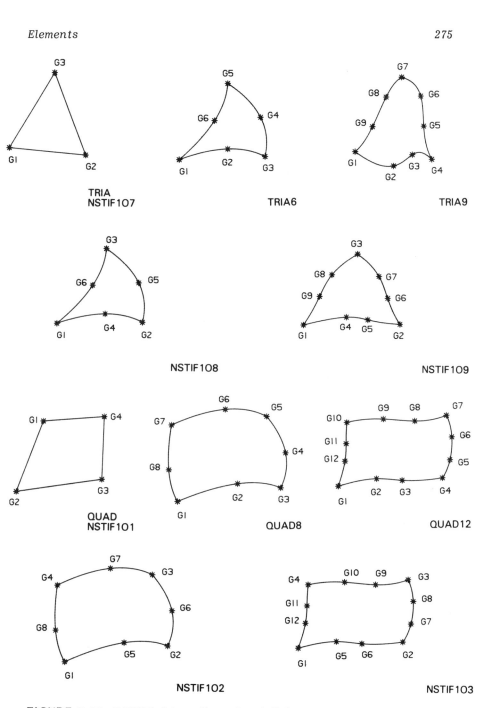

FIGURE 4.40 CADDS 4 two-dimensional finite elements.

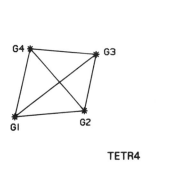

TETR4

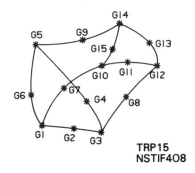

TRP15
NSTIF408

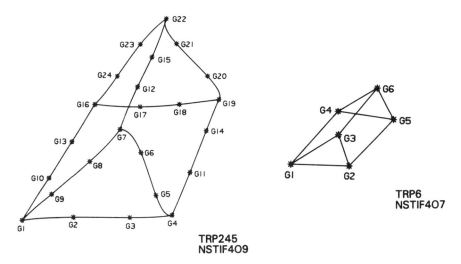

TRP245
NSTIF409

TRP6
NSTIF407

FIGURE 4.41 CADDS 4 three-dimensional finite elements.

The command INSERT ELEMENT TYPE and a space is keyed in with
the name of the particular element type keyed in after this command.
In this case, the element type is NSTIF503, which is shown in Figs.
4.39 through 4.41, or it may be the name of an element created by the
DEFINE FEMSYSTEM command. The RETURN key is pressed and if the
user does not want to use any second- or third-level modifiers, the
RETURN key should be pressed again. Then a colon is keyed in. The

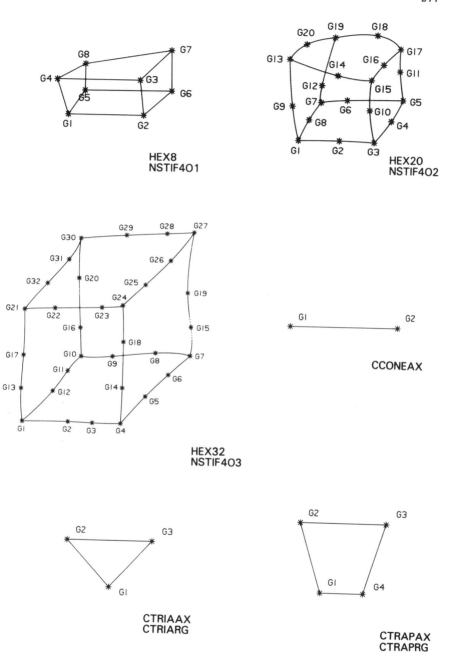

FIGURE 4.41 (Continued)

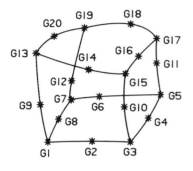

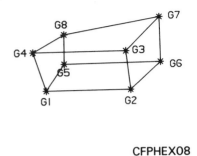

CFPHEX08

CFPHEX20

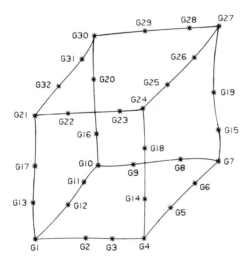

CFPHEX32

FIGURE 4.41 (Cont'd)

system prompts MODEL ent n nodes (where n is the number of grid points needed to define the type of element the user is inserting; i.e., 4). Respond by identifying the appropriate number of grid points in the order required for the element type (i.e., Tag G10-13). The RETURN key is pressed and if the user is inserting a type of element that he or she has not used previously in the part, the system dis-

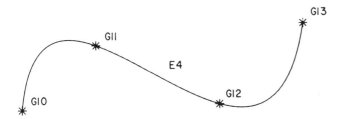

FIGURE 4.42

plays the message element type WAS ADDED TO LOCAL FILE; otherwise, no message appears.

The GENERATE ELEMENT command creates a regular pattern of elements of the same type automatically. This command is designed as a companion for GENERATE GRID POINT, although they need not be used together. The command GENERATE GRID POINT provides a number of options for grid point spacing. By filling grid point patterns created by GENERATE GRID POINT with elements created by GENERATE ELEMENT, the user can obtain smoothly graded transitions between fine and coarse element meshes or rows.

Example 2

#n# GENERATE ELEMENT SEPARATE INCREMENT 1 : MODEL
 ent TAG E1 E4 [CR] (See Fig. 4.43)

Assume that grid points 1–5 and elements 1 and 4 (which are both two-node beams) have previously been inserted in this example. The following command line will create elements 2 and 3.

Example 3

#n# GENERATE ELEMENT NEST IINCREMENT 1 JINCREMENT 3 :
 MODEL ent TAG E5 E7 E11 [CR] (see Fig. 4.44)

In this example, the bounding elements E5, E7, and E11 have previously been insertedon the grid point pattern. Since these are two-

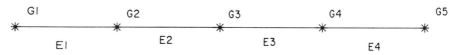

FIGURE 4.43

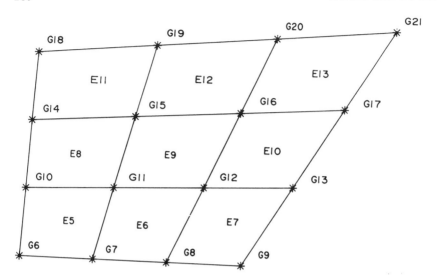

FIGURE 4.44

dimensional elements (geometric type QUAD), three bounding elements
are necessary. Note that the system interprets the I direction as lying between E5 and E7, and the J direction as lying between E5 and
E11. These directions are undefined until the user begins specifying
bounding elements. For example, the following command line produces exactly the same result as the command line above.

#n# GENERATE ELEMENT NEST IINCREMENT 3 JINCREMENT 1 :
     MODEL ent TAG E5 E11 E7 [CR]

Example 4

#n# GENERATE ELEMENT NEST IINCREMENT 1 JINCREMENT 4 :
     MODEL ent TAG E1 E4 E5 [CR] (see Fig. 4.45)

This is an example of an unusual element numbering scheme which can
be created in two steps. Since the user must use the GENERATE ELE-
MENT command twice to create the mesh, two sets of three bounding
elements are necessary. These bounding elements are E1, E4, E5
(first set) and E9, E12, E13 (second set). At this point, elements 1
through 8 are in place. The following command line completes the pattern:

#n# GENERATE ELEMENT NEST IINCREMENT 1 JINCREMENT 4 :
     Model ent TAG E9 E12 E13 [CR]

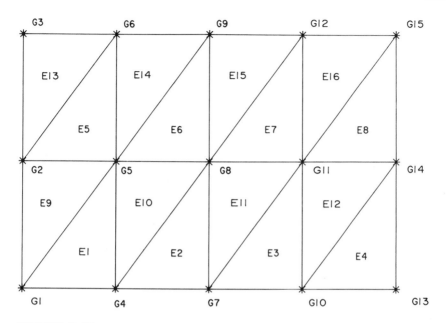

FIGURE 4.45

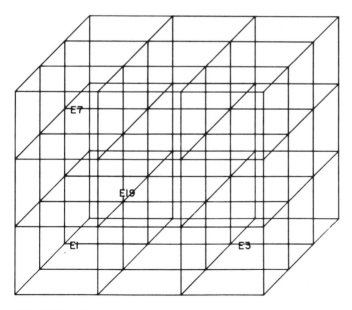

FIGURE 4.46

Example 5

#n# GENERATE ELEMENT NEST IINCREMENT 1 JINCREMENT 3
     KINCREMENT 9 : MODEL ent TAG E1 E3 E7 E19 [CR] (see Fig.
     4.46)

In this illustration of a three-dimensional mesh, no grid points or entity tags are shown except for the tags of the four bounding elements:
E1 occupies the lower left front corner, E3 occupies the lower right
front corner, E7 occupies the upper left front corner, and E19 occupies the lower left rear corner. For the purposes of this example, the
I, J, and K directions will be equated with the X, Y, and Z axes, respectively. This provides the following values for the increment modifiers: IINCREMENT is 1 (corresponding to the X-positive direction),
JINCREMENT is 3 (corresponding to the Y-positive direction), and
KINCREMENT is 9 (corresponding to the Z-negative direction).

The GENERATE MESH command creates the finite element mesh
automatically. Automatic mesh generation means the automatic crea-

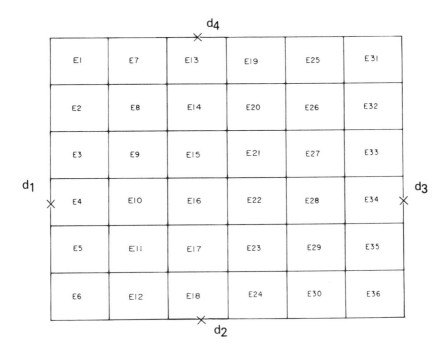

FIGURE 4.47

tion of grid points and elements. This command can considerably reduce the amount of time required to create a finite element model.

Example 6

#n# GENERATE MESH AUTOMATIC TYPE NSTIF101 [CR]
   MDESC AG90 [CR]
   ORIENT 30.0 : MODEL ent d_1; MODEL ent d_2; MODEL ent
   d_3; MODEL ent d_4 [CR]
   UNH 6 UNV 6 [CR] (see Fig. 4.47)

In this example, grid points and grid point tags are not shown for the sake of clarity. In this example, a two-dimensional region is meshed with NSTIF101 elements. Attached to the elements is an Mdescription named AG90 at an orientation angle of 30.0. Note that a semicolon separates the digitizes that identify each side of the region. Six uniformly spaced pieces are then specified for the two opposite sides of the region.

Example 7

#n# GENERATE MESH AUTOMATIC TYPE NSTIF101 SNEL 5 INCE 10:
   MODEL ent d_1; MODEL ent d_2; MODEL ent d_3; MODEL ent d_4d_5
   [CR]
   [CR]
   INPUT FOR SIDE 1, TYPE D FOR DIG, N FOR NUMBERS,
   REPLY = N [CR]
   INPUT FOR SIDE 2 TYPE D FOR DIG, N FOR NUMBERS,
   REPLY = N [CR]
   INPUT FOR SIDE 3, TYPE D FOR DIG, N FOR NUMBERS,
   REPLY = N [CR]
   INPUT FOR SIDE 4, TYPE D FOR DIG, N FOR NUMBERS,
   REPLY = N [CR]
   SIDE 1, CURVE 1 OF 1, N = 8 [CR]
   SIDE 2, CURVE 1 OF 1, N = 8 [CR]
   SIDE 1 HAS 8 PIECES — SIDE 3, CURVE 1 OF 1, N = 8 [CR]
   SIDE 2 HAS 8 PIECES — SIDE 4, CURVE 1 OF 2, N = 4 [CR]
   SIDE 2 HAS 8 PIECES — SIDE 4, CURVE 2 OF 2, N = 4 [CR]
   (see Fig. 4.48)

Grid points and grid point tags are now shown in the illustration for this example. Since the number of pieces for each pair of sides in the region is not specified in the command by the UNH and UNV modifiers, it must be specified in the Getdata. The responses to the first four prompts establish whether grid points are to be inserted or numbers are to be entered to define the number of pieces. In this case, number input is specified, and the system then requests these values.

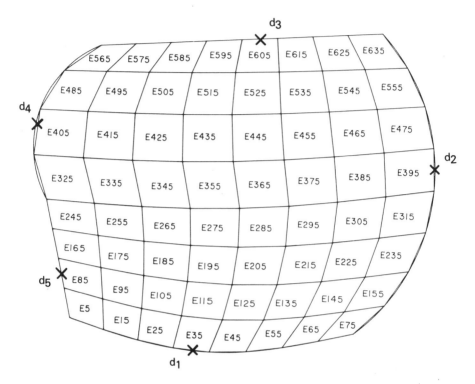

FIGURE 4.48

Example 8

#n# GENERATE MESH AUTOMATIC TYPE HEX08 SOLID SNEL 1
    INCE 1 [CR]
    INUMEL 4 JNUMEL 4 KNUMEL 3 : DIGITIZE SOLID ELEMENT
    MODEL ent TAG E100 [CR] (Fig. 4.49)

In Fig. 4.30, no grid points or grid point tags are shown (the volume
has been rotated to show its structure more clearly). Element tags
are shown only for elements lying in the corners of the cube. Assume
that the volume has previously been defined as E100 (type CFPHEX08)
and that the grid points are ordered counterclockwise starting at the
lower left front corner of the solid element.

Example 9

#n# GENERATE MESH ROTATE [CR]
    VECTOR ANGLE 30 [CR]
    GPINCREMENT 50 REPEAT 11 : MODEL ent d_1; MODEL loc d_2d_3
    [CR] (see Fig. 4.50)

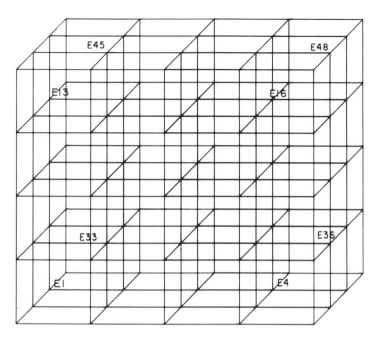

FIGURE 4.49

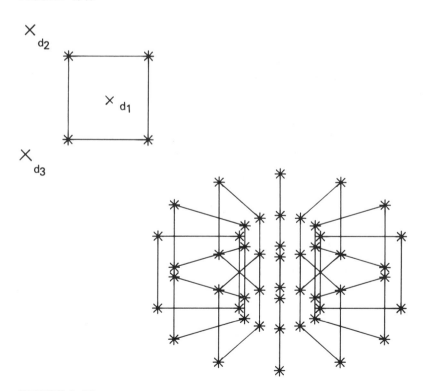

FIGURE 4.50

In this example, a single QUAD element is rotated and copied 11 times around a rotation vector. The command line specifies angular increments of 30° and a grid point increment of 50. The result has been rotated to show its structure more clearly. The first digitize (d_1) identifies the element to be rotated and copied. A semicolon separates this information from the two digitizes that define the rotation vector (d_2 and d_3). Note the use of RETURNs to separate modifier levels.

4.23.3 Editing and Deleting Elements

The user may identify elements to be edited or deleted by digitizing them near their tags or by specifying their tags in the Getdata portion of the command. The commands for editing and deleting elements are as follows: CHANGE ELEMENT, RENUMBER ELEMENT, and DELETE ENTITY.

The CHANGE ELEMENT command changes the data associated with an element. The user may change or add a reference to an element property or material description, as well as change or add attributes.

Example 10

```
#n# CHANGE ELEMENT:MODEL ent TAG E30 [CR]
    MDESCRIPTION ZN12 [CR]
    [CR]
```

In this example, element 30 becomes associated with a previously created material description named ZN12.

Example 11

```
#n# CHANGE ELEMENT:MODEL ent TAG E12-13 [CR]
    ELTYPE CORE MDESCRIPTION PB132 EPROPERT MPIPE)
    PINB 14 PINE 6 [CR]
```

Elements 12 and 13 are flagged with the ELTYPE name CORE and become associated with a material description named PB132 and a property description named MPIPE. The attribute modifiers PINB and PINE change the constraints applied to the beginning and end of the elements, respectively.

The RENUMBER ELEMENT command renumbers the elements. The existing element numbers are updated by adding a constant increment or are replaced by a new set of numbers specified in the command.

Example 12

```
#n# RENUMBER ELEMENT TO 12-22 25-28 : MODEL ent TAG E30-44
    [CR]
```

Elements 30—40 re renumbered to 12—22 and elements 41—44 are renumbered to 25—28.

 The DELETE ENTITY command is used to delete grid points and/or
elements. It removes all the information about the specified grid
points/elements from the part and updates all references and associations to maintain the integrity of the FEM part. When the user deletes
a grid point, all the elements incident on it are deleted and all references to it are updated. When the user deletes an element, only its
references are updated.

4.23.4 Controlling Element Appearance

As mentioned previously in connection with grid points, it is often
useful (and sometimes necessary) to change the appearance of elements for the sake of a clear display. The user may display or remove element tags from the screen using the SELECT TAG and ECHO
TAG commands or may blank out the elements themselves using the
BLANK ENTITY command. To redisplay blanked elements, UNBLANK
ENTITY is used. Additionally, the user may place elements on a layer
other than that used for grid points. To do this, the SELECT LAYER
or CHANGE LAYER command is used. The user may then echo the elements or remove them from the screen using the ECHO LAYER command.

4.23.5 Element Status Information

Three commands provide information about the elements in the model.
The three commands are LIST FEM, HIGHLIGHT FEM, and DISPLAY
FEMGEOMETRY.

 The LIST FEM command provides summary or detailed status information about a FEM part or about the FEM system itself.

 Example 13

a. #n# LIST FEM GPNT : MODEL ent TAG G25 [CR]

This command line provides detailed information on grid point 25.

b. #n# LIST FEM EDF [CR]
      SYSFIL [CR]
      TYPE NSTIF501 [CR]

This command line provides informaton about the NSTIF501 element type
as it is defined in DATA.FEM.DEFAULT.EDF.

 The HIGHLIGHT FEM command allows the user to visually distinguish the finite elements and grid points in a model which have a par-

ticular set of properties. Highlighting is achieved by blinking the entities' graphics or by changing their color. The user can define highlighting by element or grid point number, element type, material descriptions and element properties, grid point or element loads, and grid point constraints. The user may specify up to four properties in the command, unless one of them is a material description, in which case the maximum is three. Only entities having all the given properties will be highlighted.

   Example 14

a. #n# HIGHLIGHT FEM ELEMENT NUM 100-200 250 EPROP TYPE
   PHEX8 [CR]

The system examines elements numbered 100—200 and 250 and highlights those to which the Eproperty type PHEX8 is attached. This command relies on the default highlighting mode.

b. #n# HIGHLIGHT FEM OFF ELEMENT NUM 100-200 [CR]

This command de-highlights elements numbered 100—200 inclusive.

c. #n# HIGHLIGHT FEM OFF [CR]

Here all highlighting is turned off.
   The DISPLAY FEMGEOMETRY command displays the results of finite element analysis graphically. It can be used for element shrinking during finite element modeling.

4.24 MATERIAL DESCRIPTIONS

CADDS FEM allows the user to add information to the finite element model about the material properties attached to an element or set of elements. The user may specify up to 29 material properties, such as density, Young's modulus, and temperature characteristics.

4.24.1 Creating Material Descriptions

Material descriptions (Mdescriptions) are nongraphical CADDS entities. They have names which the user assigns when the user creates them, and the user refers to them by name at all times in FEM commands. Additionally, the user has the option of creating material descriptions before or after creating elements.
   The INSERT MATERIAL DESCRIPTION command creates a material description (Mdescription). The user must assign a different name to each Mdescription that he or she creates.

4.24.2 Editing and Deleting Material Descriptions

Use the CHANGE MDESCRIPTION and DELETE MDESCRIPTION commands to correct or delete material descriptions.

The CHANGE MATERIAL DESCRIPTION changes an existing material description. The user must include the name of the Mdescription that he or she wants to change.

Example 1 (See page 290).

#n# CHANGE MDESCRIPTION NAME ALUMINUM [CR]

This command displays the Mdescription named ALUMINUM and changes the current value of data item 24 (viscosity) to the value 0.030. The values of other data items remain unchanged.

The DELETE MATERIAL DESCRIPTION command deletes a single material description or all Mdescriptions associated with a FEM part.

Example 2

#n# DELETE MDESCRIPTION ALUMINUM [CR]

This command deletes the Mdescription having the name ALUMINUM.

4.24.3 Material Description Status Information

The user may obtain status information for a particular material description or display the association between material properties and the elements to which they are attached. The two commands are LIST MDESCRIPTION AND HIGHLIGHT FEM.

The LIST MATERIAL DESCRIPTION command displays the data values associated with material descriptions or a list of the names of all Mdescriptions associated with a part.

Example 3

#n# LIST MDESCRIPTION ALUMINUM [CR]

provides detailed information about the specified Mdescription named ALUMINUM.

HIGHLIGHT FEM allows the user to highlight various material properties in a finite element model.

4.25 ELEMENT PROPERTY DESCRIPTIONS

Another type of information associated with a FEM part is that of element property (Eproperty) descriptions, which contain structural characteristics such as mass, area, and thickness. Eproperty descriptions are classified by type: for each element type, only certain Eproperty types are appropriate.

Example 1

TO CHANGE VALUE: TYPE #, THEN VALUE: ? = LIST; OK = SAVE;
CR = EXIT

1	YOUNG'S MODULUS X	= 0.000	16	POISSON'S RATIO Z	= 0.000
2	SHEAR MODULUS X-Y	= 0.000	17	THERMAL COEFFICIENT Y	= 0.000
3	POISSON'S RATIO X	= 0.000	18	THERMAL COEFFICIENT Z	= 0.000
4	MASS DENSITY	= 0.000	19	SPECIFIC HEAT	= 0.000
5	THERMAL COEFF X	= 0.000	20	THERMAL CONDUCT X	= 0.000
6	REF TEMPERATURE	= 0.000	21	THERMAL CONDUCT Y	= 0.000
7	DAMPING COEFFICIENT	= 0.000	22	THERMAL CONDUCT Z	= 0.000
8	TENSION LIMIT	= 0.000	23	CONVECTION	= 0.000
9	COMPRESSION LIMIT	= 0.000	24	VISCOSITY	= 0.020
10	SHEAR LIMIT	= 0.000	25	RESISTIVITY X	= 0.000
11	YOUNG'S MODULUS Y	= 0.000	26	RESISTIVITY Y	= 0.000
12	YOUNG'S MODULUS Z	= 0.000	27	RESISTIVITY Z	= 0.000
13	SHEAR MODULUS Y-Z	= 0.000	28	EMISSIVITY	= 0.5700
14	SHEAR MODULUS X-Z	= 0.000	29	COEFF OF FRICTION	= 0.000
15	POISSON'S RATIO Y	= 0.000			

```
24  [CR]
24  VISCOSITY                = 0.02000
=   0.03 [CR]
    [CR]
```

1	YOUNG'S MODULUS X	= 0.000	16	POISSON'S RATIO Z	= 0.000
2	SHEAR MODULUS X-Y	= 0.000	17	THERMAL COEFFICIENT Y	= 0.000
3	POISSON'S RATIO X	= 0.000	18	THERMAL COEFFICIENT Z	= 0.000
4	MASS DENSITY	= 0.000	19	SPECIFIC HEAT	= 0.000
5	THERMAL COEFF X	= 0.000	20	THERMAL CONDUCT X	= 0.000
6	REF TEMPERATURE	= 0.000	21	THERMAL CONDUCT Y	= 0.000
7	DAMPING COEFFICIENT	= 0.000	22	THERMAL CONDUCT Z	= 0.000
8	TENSION LIMIT	= 0.000	23	CONVECTION	= 0.000
9	COMPRESSION LIMIT	= 0.000	24	VISCOSITY	= 0.030
10	SHEAR LIMIT	= 0.000	25	RESISTIVITY X	= 0.000
11	YOUNG'S MODULUS Y	= 0.000	26	RESISTIVITY Y	= 0.000
12	YOUNG'S MODULUS Z	= 0.000	27	RESISTIVITY Z	= 0.000
13	SHEAR MODULUS Y-Z	= 0.000	28	EMISSIVITY	= 0.5700
14	SHEAR MODULUS X-Z	= 0.000	29	COEFF OF FRICTION	= 0.000
15	POISSON'S RATIO Y	= 0.000			

TYPE OK TO STORE DATA VALUES
OK [CR]

4.25.1 Creating Element Property Descriptions

Like material descriptions, Eproperty descriptions are nongraphical CADDS entities to which the user assigns names as the user creates them. Each element may refer to one property description, although the description need not be unique to that element. All references to property descriptions in a FEM command are by name, and the Eproperty description may be created before or after elements are created.

The INSERT ELEMENT PROPERTY command creates an element property (Eproperty) description. The user must assign a different name to each Eproperty created.

Example 1

a. #n# INSERT EPROPERTY NAME ALUMINUM TYPE PPIPE [CR]

TO CHANGE VALUE: TYPE #, THEN VALUE; ? = LIST; OK = SAVE; CR = EXIT
1. OUTSIDE DIAMETER OF PIPE   = 0.000
2. PIPE WALL THICKNESS   = 0.000
3. MASS PER UNIT LENGTH   = 0.000
4. INTERNAL PRESSURE   = 0.000

2   [CR]
2. PIPE WALL THICKNESS = 0.000
=   0.7 [CR]
3   [CR]
3. MASS PER UNIT LENGTH = 0.000
=   1.02 [CR]
[CR]
1. OUTSIDE DIAMETER OF PIPE   = 0.000
2. PIPE WALL THICKNESS   = 0.700
3. MASS PER UNIT LENGTH   = 1.020
4. INTERNAL PRESSURE   = 0.000
TYPE OK TO STORE DATA VALUES

OK [CR]

This command creates an Eproperty type PPIPE named ALUMINUM and assigns the value 0.7 to data item 2 (pipe wall thickness), and the value 1.02 to data item 3 (mass per unit length). All other data items remain their original values of 0.000.

b. #n# INSERT MDESCRIPTION NAME ALUMINUM [CR]

TO CHANGE VALUE: TYPE #, THEN VALUE; ? = LIST; OK = SAVE; CR = EXIT

1	YOUNG'S MODULUS X	= 0.000	16	POISSON'S RATIO Z	= 0.000
2	SHEAR MODULUS X-Y	= 0.000	17	THERMAL COEFFICIENT Y	= 0.000
3	POISSON'S RATIO X	= 0.000	18	THERMAL COEFFICIENT Z	= 0.000
4	MASS DENSITY	= 0.000	19	SPECIFIC HEAT	= 0.000
5	THERMAL COEFF X	= 0.000	20	THERMAL CONDUCT X	= 0.000
6	REF TEMPERATURE	= 0.000	21	THERMAL CONDUCT Y	= 0.000
7	DAMPING COEFFICIENT	= 0.000	22	THERMAL CONDUCT Z	= 0.000
8	TENSION LIMIT	= 0.000	23	CONVECTION	= 0.000
9	COMPRESSION LIMIT	= 0.000	24	VISCOSITY	= 0.000
10	SHEAR LIMIT	= 0.000	25	RESISTIVITY X	= 0.000
11	YOUNG'S MODULUS Y	= 0.000	26	RESISTIVITY Y	= 0.000
12	YOUNG'S MODULUS Z	= 0.000	27	RESISTIVITY Z	= 0.000
13	SHEAR MODULUS Y-Z	= 0.000	28	EMISSIVITY	= 0.000
14	SHEAR MODULUS X-Z	= 0.000	29	COEFF OF FRICTION	= 0.000
15	POISSON'S RATIO Y	= 0.000			

```
24  [CR]
24.  VISCOSITY              = 0.000
=   0.02 [CR]
28  [CR]
28.  EMISSIVITY             = 0.000
=   0.57 [CR]
[CR]
```

This command creates an Mdescription named ALUMINUM and assigns the vlaue of 0.02 to data item 24, which is viscosity, and the value of 0.57 to data item 28, which is emissivity. All other data items retain their original values of 0.000.

4.25.2 Editing and Deleting Element Property Descriptions

Use the CHANGE EPROPERTY AND DELETE EPROPERTY commands to correct or delete Eproperties.

The CHANGE ELEMENT PROPERTY changes an existing Eproperty. The user must include the name of the Eproperty that is to be changed.

Example 2

#n# CHANGE EPROPERTY NAME ALUMINUM [CR]

```
TO CHANGE VALUE:  TYPE #, THEN VALUE: ? = LIST; OK = SAVE;
CR = EXIT
1.  OUTSIDE DIAMETER OF PIPE    = 0.000
2.  PIPE WALL THICKNESS         = 0.7000
3.  MASS PER UNIT LENGTH        = 1.020
4.  INTERNAL PRESSURE           = 0.000
```

2   [CR]
2. PIPE WALL THICKNESS = 0.700
=   0.6 [CR]
[CR]

TO CHANGE VALUE: TYPE #, THEN VALUE: ? = LIST; OK = SAVE;
CR = EXIT
1. OUTSIDE DIAMETER OF PIPE   = 0.000
2. PIPE WALL THICKNESS   = 0.600
3. MASS PER UNIT LENGTH   = 1.0200
4. INTERNAL PRESSURE   = 0.000
TYPE OK TO STORE DATA VALUES

OK   [CR]

This command displays the previously created Eproperty named
ALUMINUM (type PPIPE) and changes the current value of data item
2 (pipe wall thickness) to the value 0.6. The values of all other data
items remain unchanged.

The DELETE ELEMENT PROPERTY command deletes a single Eproperty or all Eproperties associated with a FEM part.

Example 3

#n# DELETE EPROPERTY NAME AU1 [CR]

This command deltes an Eproperty named AU1.

4.25.3 Element Property Description Status Information

LIST EPROPERTY and HIGHLIGHT FEM commands are used to obtain
information about Eproperties associated with the FEM part.

The LIST ELEMENT PROPERTY command displays the data values
associated with Eproperties or a list of the names of all Eproperties
associated with a part.

Example 4

#n# LIST EPROPERTY NAMSONLY [CR]

This command displays a list of the names of all Eproperties associated
with a FEM part.

The HIGHLIGHT FEM enables the user to highlight the model to
show the association between Eproperties and the elements to which
they are attached.

4.26 LOADS

The definition of the external environment of the finite element model
is one kind of data used as input to FEA programs. External loads
constitute one aspect of this environment definition (constraints constitute another aspect).

In CADDS FEM the user can define the following loads: forces and
moments, displacements and rotations, temperatures, pressure, and
temperature gradients for grid point loads. Pressure loads on surface edges, traction loads on surface edges, body forces (through volume), and temperature/temperature gradient can be defined for element loads.

4.26.1 Creating Load Sets

The INSERT LOAD command creates a load set containing one or more
grid point or element loads. The load set is identified by its set name
and the individual loads in the set by their ID numbers. The user may
use this command wither for inserting individual loads at the initial
creation of the load set or for updating values in an existing load set.

Example 5

a. #n# INSERT LOAD SET LS1 STATIC [CR]
        GPLOAD GLOBAL NUMBER 1 [CR]
        FORCE Y-50 : MODEL ent TAG G1110-1111 [CR]

This command line creates a static load set named LS1 which contains
one grid point load instance whose ID is 1. The load is applied with
respect to the global coordinate system and represents a force of Y-50
applied to grid points G1110 and G1111.

b. #n# INSERT LOAD SET LOAD1 [CR]
        GPLOAD NUMBER 2 [CR]
        FORCE X1.2 Y2.0 [CR]
        MOMENT X1.0 Z2.0 [CR]
        PRESSURE 3.3 : MODEL ent $d_1 \cdots d_n$ [CR]

Here a load set named LOAD1 is created. The load instance ID is assigned explicitly (NUMBER 2) and contains the following values:

        Force X = 1.2
        Force Y = 2.0
        Moment X = 1.0
        Moment Z = 2.0
        Pressure = 3.3

4.26.2 Editing and Deleting Loads

The two commands used to correct or delete loads are CHANGE LOAD and DELETE LOAD. The CHANGE LOAD command is used to change the load set name or load instance ID numbers. It will also add or delete a list of grid points or elements associated with an existing load. (To change load vlaues in an existing load set, use INSERT LOAD.)

Example 2

a. #n# CHANGE LOAD SET LS3 NUMBER5 : MODEL ent TAG G15-20 [CR]

The result of this command is that grid points 15—20 are added to the load number 5 contained in the set named LS3.

b. #n# CHANGE LOAD SET LS3 NEWNAME LS4 [CR]

The result of this command is that LS3 is renamed to LS4.

The DELETE LOAD command deletes a single grid point or element load, all loads in a load set, or all sets associated with a FEM part.

Example 3

#n# DELETE LOAD NAME LS6 NUMBER 6 [CR]

This command lie deletes load number 6 from the set named LS6.

4.26.3 Load Status Information

The LIST LOAD command provide status information about the loads applied to a finite element model.

Example 4

LIST LOAD NAME LS20 [CR]

This command line displays information about all loads in the load set named LS20.

4.27 CONSTRAINTS

In the simulation of the environment of a finite element part, constraints indicate the degrees of freedom of a grid point which are constrained. The user defines the boundary conditions of the finite element model by specifying single-point constraint sets for the grid points. A single-point constraint set (SPCset) defines a group of grid points to which the same constraint instances are applied. These instances allow the finite element analysis program to change the boundary conditions on a structure by enabling or disabling the single-point constraint sets. The user may also create sets of multiple-point constraints: these enable the user to define the interaction of constraint conditions for a group of grid points.

4.27.1 Creating Single-Point Sets

Single-point constraint sets are referenced by names that the user assigns to them. Sets consist of constraint instances which, like load instances, are assigned integer IDs. The user may create SPCsets at any time during the modeling process.

The INSERT SINGLE POINT CONSTRAINT SET command creates a single-point constraint set. The user must supply a name for the set and the constraint values to be used for each constraint instance. The user then digitizes the grid points to be constrained in the instance. The constraint specifies the degrees of freedom, which are constrained and which are represented by combinations of the digits 1 through 6. The digits and the constraints they represent are as follows:

1 = X translation   4 = X rotation
2 = Y translation   5 = Y rotation
3 = Z translation   6 = Z rotation

Example 1

#n# INSERT SPCSET NAME SETI [CR]
    SPCAX [CR]
    PSPC 123 HARMONIC 1 PSPC 234 HARMONIC 2 PSPC 234
    HARMONIC 3 : CONSTRAINT INSTANCE NO. 1 MODEL ent :
    $d_1 d_2$ [CR]

This example specifies the following boundary conditions on the two digitized ring grid points.

Harmonic	Constraint value
1	123
2	234
3	234

4.27.2 Editing and Deleting Single-Point Constraint Sets

The CHANGE SPCSET and DELETE SPCSET commands correct or delete SPCsets.

The CHANGE SINGLE POINT CONSTRAINT SET command modifies an existing single-point constraint set. The user can either add grid points to a specified constraint instance or remove grid points from the constraint instance, but not both. The user may also change the values of the constraint instance. If the specified constraint instance does not exist, a new one will be created.

Example 2

#n# CHANGE SPCSET NAME SET1 [CR]
    NUMBER 5 PSPC 135 : MODEL ent TAG G11-37 [CR]

Grid points 11–37 are added to constraint instance number 5 (which is assigned the value 135) in the SPCset named SET1.

The DELETE SINGLE POINT CONSTRAINT SET command deletes one (or all) single-point constraint set(s) or a specified constraint instance from a constraint. The user must specify the name of the set that is to be deleted or changed.

Example 3

#n# DELETES SPCSET NAME LS3 NUMBER 5 [CR]

This command line deletes constraint instance 5 from the SPCset named LS3.

4.27.3 Single-Point Constraint Set Status Information

The List SPCSET command enables the user to list the names of single-point constraint sets, the values in a set, the values for individual instances, grid point numbers, as well as complete information on all sets associated with a part.

Example 4

#n# LIST SPCSET NAME LS4 PNTS [CR]

This command provides information about the SPCset named LS4 and its associated grid points.

4.28 PREPARING FEM DATA FOR FINITE ELEMENT ANALYSIS

Once the finite element model is complete, the user is ready to create the necessary files which the finite element analysis (FEA) program uses to analyze the model. The PUT FEM command creates these files, which are made up of text card images describing the entities associated with the finite element model. No job control language (JCL) cards are generated, but the user may create and add them using one of the Computervision text editors.

The following finite element analysis programs are supported by PUT FEM: (1) ANSYS Rev. 4, (2) NASTRAN, (3) NASTRAX (axisymmetric NASTRAN), (4) SAP Rev. 5, (5) STRUDL, (6) SUPERB, and (7) CVFEM.

Example 1

a. #n# PUT FEM PRINT [CR]

DATA PUTFEM EXISTS
TYPE OK TO OVERWRITE
OK [CR]

PROCESSING DATA

CREATING TEXT FILE

This command line creates the cards according to NASTRAN format (default) as defined in the file called DATA.FEMLANG.SECURE.NAS-TRAN; stores the cards in the text file called DATA.&BCD.PUTFEM (default name; an old version of this file is overwritten); displays the cards on the screen as they are created; and outputs information related to the selected FEA language.

b. #n# PUT FEM PRINT FEMLANG SUPERB [CR]

PROCESSING DATA

CREATING TEXT FILE

ANSYS-compatible cards will be placed into the file named DATA.& BCD.FEM2. No card images will be displayed.

d. #n# PUT FEM FEMLANG CVFEM TEXTFILE CARDIMS.CVFEM [CR]

PROCESSING DATA

CREATING TEXT FILE

CVFEM-compatible card images will be placed into the file CARDIMS.& BDC.CVFEM. Card images will not be displayed as they are created.

4.29 GRAPHIC DISPLAY OF FEA RESULTS

CADDS FEM allows to display the results of finite element analysis graphically at the user's workstation:

1. The DISPLAY FEMGEOMETRY command enables the user to display a deflected shape (static deformation). With this command the user may also view animation of the cycle deformation of a structure at different frequencies, or animation of a transient deformation of a model due to transient loading.
2. The SAVE ANIMATION and STORE ANIMATION commands enable the user to store and redisplay animation frames created by DISPLAY FEMGEOMETRY.

Additional commands enable the user to display such results as stress or temperature characteristics:

3. GENERATE CONTOUR, FILL CONTOUR, and DISPLAY CONTOUR create contour displays of FEA results.
4. FILL ELEMENTS and COLOR FEM create displays in which grid points or elements are colored to show the results of analysis.

4.30 FINITE ELEMENT ANALYSIS

After the finite element model is created from the graphics model, this finite element model is analyzed by using one of the analysis programs. The following examples illustrate the results obtained from such analysis programs.

Example 1: Static Analysis

The chemical processing tank shown in Fig. 4.51 is to be analyzed using SUPERB. For the purpose of simply illustrating use of SUPERB, the horizontal semicircular flat plate will be analyzed as a separate component. Since the boundary conditions and loading conditions to be applied to the plate are symmetric about the X-Z plane, one-fourth of the plate will be modeled and symmetric boundary conditions will be applied to nodes on the X axis. The outer circular edge of the plate will be considered to be clamped. A constant internal pressure of 200 psi will be applied to the top of the plate. The parabolic thin shell element Fig. 4.52 will be used to represent the plate. Plate thickness = 0.75 in.

Figure 4.53 shows a plot of the nodal points making up the finite element model, and Fig. 4.54 shows the numbering sequence for the various nodal points. The six finite elements are shown in Fig. 4.55

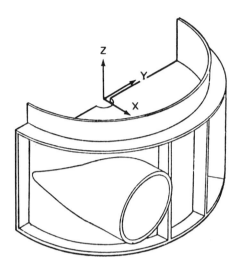

FIGURE 4.51 Pressure vessel.

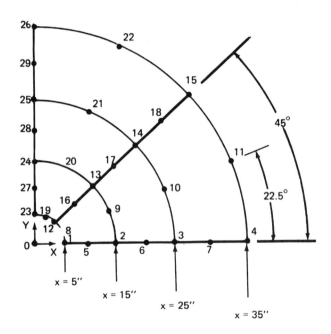

FIGURE 4.52

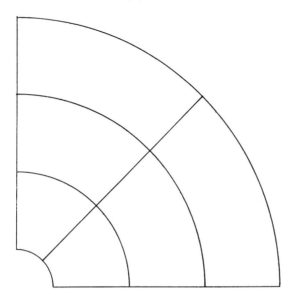

FIGURE 4.53 Static analysis example: pressure vessel lid, plot 1.

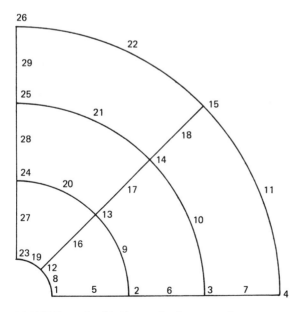

FIGURE 4.54 Static analysis example: pressure vessel lid, plot 2.

STATIC ANALYSIS EXAMPLE - PRESSURE VESSEL LID

LOADING CASE   1 (SEQUENCE 1) INTERNAL PRESSURE LOADING
  *** DISPLACEMENT SOLUTION ***

NODE	UX	UY	UZ	ROTX	ROTY	ROTZ
1	0.00000E+00	0.00000E+00	5.42124E-01	0.00000E+00	1.89958E-02	0.00000E+00
2	0.00000E+00	0.00000E+00	3.45980E-01	0.00000E+00	2.18487E-02	0.00000E+00
3	0.00000E+00	0.00000E+00	1.22878E-01	0.00000E+00	2.01738E-02	0.00000E+00
4	0.00000E+00	0.00000E+00	4.47835E-01	0.00000E+00	0.00000E+00	0.00000E+00
5	0.00000E+00	0.00000E+00	2.32336E-01	0.00000E+00	1.92123E-02	0.00000E+00
6	0.00000E+00	0.00000E+00	3.61993E-02	0.00000E+00	2.28732E-02	0.00000E+00
7	0.00000E+00	0.00000E+00	5.42661E-01	-7.21707E-03	1.74236E-02	0.00000E+00
8	0.00000E+00	0.00000E+00	3.45666E-01	-8.36500E-03	2.01949E-02	0.00000E+00
9	0.00000E+00	0.00000E+00	1.22336E-01	-7.78097E-03	1.87849E-02	0.00000E+00
10	0.00000E+00	0.00000E+00	0.00000E+00	0.00000E+00	0.00000E+00	0.00000E+00
11	0.00000E+00	0.00000E+00	5.42124E-01	-1.34321E-02	1.34321E-02	0.00000E+00
12	0.00000E+00	0.00000E+00	3.45980E-01	-1.54494E-02	1.54494E-02	0.00000E+00
13	0.00000E+00	0.00000E+00	1.22878E-01	-1.42650E-02	1.42650E-02	0.00000E+00
14	0.00000E+00	0.00000E+00	0.00000E+00	0.00000E+00	0.00000E+00	0.00000E+00
15	0.00000E+00	0.00000E+00	4.47835E-01	-1.35851E-02	1.35851E-02	0.00000E+00
16	0.00000E+00	0.00000E+00	2.32336E-01	-1.61738E-02	1.61738E-02	0.00000E+00
17	0.00000E+00	0.00000E+00	3.61993E-02	-1.74236E-02	9.32018E-03	0.00000E+00
18	0.00000E+00	0.00000E+00	5.42661E-01	-2.01949E-02	7.21707E-03	0.00000E+00
19	0.00000E+00	0.00000E+00	3.45666E-01	-1.87849E-02	8.36500E-03	0.00000E+00
20	0.00000E+00	0.00000E+00	1.22336E-01	0.00000E+00	7.78097E-03	0.00000E+00
21	0.00000E+00	0.00000E+00	0.00000E+00	0.00000E+00	0.00000E+00	0.00000E+00
22	0.00000E+00	0.00000E+00	5.42124E-01	-1.89958E-02	0.00000E+00	0.00000E+00
23	0.00000E+00	0.00000E+00	3.45980E-01	-2.18487E-02	0.00000E+00	0.00000E+00
24	0.00000E+00	0.00000E+00	1.22878E-01	-2.01738E-02	0.00000E+00	0.00000E+00
25	0.00000E+00	0.00000E+00	4.47835E-01	0.00000E+00	0.00000E+00	0.00000E+00
26	0.00000E+00	0.00000E+00	0.00000E+00	-1.92123E-02	0.00000E+00	0.00000E+00
27	0.00000E+00	0.00000E+00	2.32336E-01	-2.28732E-02	0.00000E+00	0.00000E+00
28	0.00000E+00	0.00000E+00	3.61993E-02	-1.31807E-02	0.00000E+00	0.00000E+00

****** MAXIMUM DISPLACEMENTS ******

	NODE	VALUE
	0	0.00000E+00
	0	0.00000E+00
	1	5.42124E-01
	28	-2.28732E-02
	6	2.28733E-02
	0	0.00000E+00

STATIC ANALYSIS EXAMPLE - PRESSURE VESSEL LID

LOADING CASE 1 (SEQUENCE 1) INTERNAL PRESSURE LOADING

AVERAGE NODAL STRESS AT TOP, MIDDLE AND BOTTOM OF THIN PLATE AND SHELL ELEMENTS

```
**********************************************************************
*AVERAGED STRESSES AT NODES WHERE GEOMETRIC DISCONTINUITIES EXIST IN SURFACE*
* SLOPES OF ADJOINING ELEMENTS ARE INVALID AND ARE NOT LISTED
**********************************************************************
```

NODE	*STS*	STRESS-X	STRESS-XY	STRESS-Y	STRESS-XZ	STRESS-YZ	STRESS-Z	PRINCIPAL	STRESSES	PRINCIPAL	VON MISES
1	L-STS	1.3826E+04	-3164.	7.8000E+04	0.0000E+00	0.0000E+00	0.0000E+00	1.3670E+04	0.0000E+00	7.8155E+04	7.2296E+04
		0.0000E+00	0.0000E+00	0.0000E+00	-69.23	-98.95	0.0000E+00	-120.8	0.0000E+00	-120.8	209.2
		-1.3826E+04	3164.	-7.8000E+04	0.0000E+00	0.0000E+00	0.0000E+00	-7.8155E+04	0.0000E+00	-1.3670E+04	7.2296E+04
2	L-STS	2.5047E+04	-649.8	3.8796E+04	0.0000E+00	0.0000E+00	0.0000E+00	2.5017E+04	0.0000E+00	3.8826E+04	3.4088E+04
		0.0000E+00	0.0000E+00	0.0000E+00	-1409.	-51.90	0.0000E+00	-1410.	0.0000E+00	1410.	2441.
		-2.5047E+04	649.8	-3.8796E+04	0.0000E+00	0.0000E+00	0.0000E+00	-3.8826E+04	0.0000E+00	-2.5017E+04	3.4088E+04
3	L-STS	-1.7571E+04	-1526.	1.6493E+04	0.0000E+00	0.0000E+00	0.0000E+00	1.6561E+04	0.0000E+00	1.7640E+04	2.9623E+04
		0.0000E+00	0.0000E+00	0.0000E+00	-2559.	114.7	0.0000E+00	2561.	0.0000E+00	-2561.	4436.
		1.7571E+04	1526.	-1.6493E+04	0.0000E+00	0.0000E+00	0.0000E+00	1.7640E+04	0.0000E+00	-1.6561E+04	2.9623E+04
4	L-STS	-8.3773E+04	-2973.	-1.8664E+04	0.0000E+00	0.0000E+00	0.0000E+00	-1.8528E+04	0.0000E+00	-8.3909E+04	7.6350E+04
		0.0000E+00	0.0000E+00	0.0000E+00	-3484.	-466.3	0.0000E+00	3515	0.0000E+00	-3515	6688.
		8.3773E+04	2973.	1.8664E+04	0.0000E+00	0.0000E+00	0.0000E+00	8.3909E+04	0.0000E+00	1.8528E+04	7.6350E+04
5	L-STS	2.0329E+04	-1693.	5.7064E+04	0.0000E+00	0.0000E+00	0.0000E+00	5.7142E+04	0.0000E+00	2.0251E+04	5.0181E+04
		0.0000E+00	0.0000E+00	0.0000E+00	-774.3	34.47	0.0000E+00	775.1	0.0000E+00	-775.1	1343
		-2.0329E+04	1693.	-5.7064E+04	0.0000E+00	0.0000E+00	0.0000E+00	-2.0251E+04	0.0000E+00	-5.7142E+04	5.0181E+04
6	L-STS	2660.	-1190.	2.8888E+04	0.0000E+00	0.0000E+00	0.0000E+00	2.8942E+04	0.0000E+00	2606.	2.7731E+04
		0.0000E+00	0.0000E+00	0.0000E+00	-1954.	-59.31	0.0000E+00	1955.	0.0000E+00	-1955.	3386.
		-2660.	1190.	-2.8888E+04	0.0000E+00	0.0000E+00	0.0000E+00	-2606.	0.0000E+00	-2.8942E+04	2.7731E+04
7	L-STS	-5.0486E+04	-2362.	-996.3	0.0000E+00	0.0000E+00	0.0000E+00	-883.8	0.0000E+00	-5.0599E+04	5.0163E+04
		0.0000E+00	0.0000E+00	0.0000E+00	-3016.	-195.0	0.0000E+00	3022.	0.0000E+00	-3022.	5234.
		5.0486E+04	2362.	996.3	0.0000E+00	0.0000E+00	0.0000E+00	5.0599E+04	0.0000E+00	883.8	5.0163E+04
8	L-STS	2.8288E+04	-1.7625E+04	6.3538E+04	0.0000E+00	0.0000E+00	0.0000E+00	2.0987E+04	0.0000E+00	7.0839E+04	6.3023E+04
		0.0000E+00	0.0000E+00	0.0000E+00	-94.07	-29.97	0.0000E+00	-101.8	0.0000E+00	101.8	176.4
		-2.8288E+04	1.7625E+04	-6.3538E+04	0.0000E+00	0.0000E+00	0.0000E+00	-7.0839E+04	0.0000E+00	-2.0987E+04	3.3225E+04

NODE	*STS*	STRESS-X	STRESS-XY	STRESS-Y	STRESS-XZ	STRESS-YZ	STRESS-Z	******PRINCIPAL STRESSES*****			VON MISES
9	L-STS	2.8160E+04	-3762.	3.5683E+04	0.0000E+00	0.0000E+00	0.0000E+00	3.7242E+04	2.6601E+04	0.0000E+00	3.3225E+04
	L-STS	0.0000E+00	0.0000E+00	0.0000E+00	-1221.	-505.6	0.0000E+00	1321.	-1321.	0.0000E+00	2288.
	L-STS	-2.8160E+04	3762.	-3.5683E+04	0.0000E+00	0.0000E+00	0.0000E+00	-2.6601E+04	-3.7242E+04	0.0000E+00	3.3225E+04
10	L-STS	-9819.	-9279.	8740.	0.0000E+00	0.0000E+00	0.0000E+00	1.2583E+04	-1.3662E+04	0.0000E+00	2.2736E+04
	L-STS	0.0000E+00	0.0000E+00	0.0000E+00	-2144.	-887.9	0.0000E+00	2320.	-2320.	0.0000E+00	4019.
	L-STS	9819.	9279.	-8740.	0.0000E+00	0.0000E+00	0.0000E+00	1.3662E+04	-1.2583E+04	0.0000E+00	2.2736E+04
11	L-STS	-6.8983E+04	-1.7764E+04	-3.3455E+04	0.0000E+00	0.0000E+00	0.0000E+00	-2.6096E+04	-7.6341E+04	0.0000E+00	6.7206E+04
	L-STS	0.0000E+00	0.0000E+00	0.0000E+00	-3139.	-1300.	0.0000E+00	3397.	-3397.	0.0000E+00	5884.
	L-STS	6.8983E+04	1.7764E+04	3.3455E+04	0.0000E+00	0.0000E+00	0.0000E+00	7.6341E+04	2.6096E+04	0.0000E+00	6.7206E+04
12	L-STS	4.5913E+04	-3.2087E+04	4.5913E+04	0.0000E+00	0.0000E+00	0.0000E+00	7.8000E+04	1.3826E+04	0.0000E+00	7.2088E+04
	L-STS	0.0000E+00	0.0000E+00	0.0000E+00	-48.95	-48.95	0.0000E+00	69.23	-69.23	0.0000E+00	119.9
	L-STS	-4.5913E+04	3.2087E+04	-4.5913E+04	0.0000E+00	0.0000E+00	0.0000E+00	-1.3826E+04	-7.8000E+04	0.0000E+00	7.2088E+04
13	L-STS	3.1922E+04	-6874.	3.1922E+04	0.0000E+00	0.0000E+00	0.0000E+00	3.8796E+04	2.5047E+04	0.0000E+00	3.4070E+04
	L-STS	0.0000E+00	0.0000E+00	0.0000E+00	-996.0	-996.0	0.0000E+00	1409.	-1409.	0.0000E+00	2440.
	L-STS	-3.1922E+04	6874.	-3.1922E+04	0.0000E+00	0.0000E+00	0.0000E+00	-2.5047E+04	-3.8796E+04	0.0000E+00	3.4070E+04
14	L-STS	-539.5	-1.7032E+04	-539.5	0.0000E+00	0.0000E+00	0.0000E+00	1.6493E+04	-1.7571E+04	0.0000E+00	2.9505E+04
	L-STS	0.0000E+00	0.0000E+00	0.0000E+00	-1809.	-1809.	0.0000E+00	2559.	-2559.	0.0000E+00	4432.
	L-STS	539.5	1.7032E+04	539.5	0.0000E+00	0.0000E+00	0.0000E+00	1.7571E+04	-1.6493E+04	0.0000E+00	2.9505E+04
15	L-STS	-5.1219E+04	-3.2555E+04	-5.1219E+04	0.0000E+00	0.0000E+00	0.0000E+00	-1.8664E+04	-8.3773E+04	0.0000E+00	7.6176E+04
	L-STS	0.0000E+00	0.0000E+00	0.0000E+00	-2464.	-2464.	0.0000E+00	3484.	-3484.	0.0000E+00	6035.
	L-STS	5.1219E+04	3.2555E+04	5.1219E+04	0.0000E+00	0.0000E+00	0.0000E+00	8.3773E+04	1.8664E+04	0.0000E+00	7.6176E+04
16	L-STS	3.8696E+04	-1.8368E+04	3.8696E+04	0.0000E+00	0.0000E+00	0.0000E+00	5.7064E+04	2.0329E+04	0.0000E+00	5.0095E+04
	L-STS	0.0000E+00	0.0000E+00	0.0000E+00	-547.5	-547.5	0.0000E+00	774.3	-774.3	0.0000E+00	1341.
	L-STS	-3.8696E+04	1.8368E+04	-3.8696E+04	0.0000E+00	0.0000E+00	0.0000E+00	-2.0329E+04	-5.7064E+04	0.0000E+00	5.0095E+04
17	L-STS	1.5774E+04	-1.3114E+04	1.5774E+04	0.0000E+00	0.0000E+00	0.0000E+00	2.8888E+04	2660.	0.0000E+00	2.7655E+04
	L-STS	0.0000E+00	0.0000E+00	0.0000E+00	-1382.	-1382.	0.0000E+00	1954.	-1954.	0.0000E+00	3384.
	L-STS	-1.5774E+04	1.3114E+04	-1.5774E+04	0.0000E+00	0.0000E+00	0.0000E+00	-2660.	-2.8888E+04	0.0000E+00	2.7655E+04
18	L-STS	-2.5741E+04	-2.4745E+04	-2.5741E+04	0.0000E+00	0.0000E+00	0.0000E+00	-996.3	-5.0486E+04	0.0000E+00	4.9996E+04
	L-STS	0.0000E+00	0.0000E+00	0.0000E+00	-2132.	-2132.	0.0000E+00	3016.	-3016.	0.0000E+00	5223.
	L-STS	2.5741E+04	2.4745E+04	2.5741E+04	0.0000E+00	0.0000E+00	0.0000E+00	5.0486E+04	996.3	0.0000E+00	4.9996E+04

NODE	*STS*	STRESS-X	STRESS-XY	STRESS-Y	STRESS-XZ	STRESS-YZ	STRESS-Z	******PRINCIPAL	STRESSES******		VON MISES
19	L-STS	6.3538E+04	-1.7625E+04	2.8288E+04	0.0000E+00	0.0000E+00	0.0000E+00	7.0839E+04	2.0987E+04	0.0000E+00	6.3023E+04
		0.0000E+04	0.0000E+00	0.0000E+04	-38.97	-94.07	0.0000E+00	1018	-101.8	0.0000E+00	176.4
		-6.3538E+04	1.7625E+04	-2.8288E+04	0.0000E+00	0.0000E+00	0.0000E+00	-2.0987E+04	-7.0839E+04	0.0000E+00	6.3023E+04
20	L-STS	3.5683E+04	-3762.	2.8160E+04	0.0000E+00	0.0000E+00	0.0000E+00	3.7242E+04	2.6601E+04	0.0000E+00	3.3225E+04
		0.0000E+00	0.0000E+00	0.0000E+00	-505.6	-1221.	0.0000E+00	1321.	-1321.	0.0000E+00	2288.
		-3.5683E+04	3762.	-2.8160E+04	0.0000E+00	0.0000E+00	0.0000E+00	-3.72442E+04	-3.72442E+04	0.0000E+00	3.3225E+04
21	L-STS	8740.	-9279.	-9819.	0.0000E+00	0.0000E+0	0.0000E+00	1.2583E+04	-1.3662E+04	0.0000E+00	2.2736E+04
		0.0000E+00	0.0000E+00	0.0000E+00	-887.9	-7144.	0.0000E+00	2320.	-2320.	0.0000E+00	4019.
		-8740.	9279.	9819.	0.0000E+00	0.0000E+00	0.0000E+00	1.3662E+04	1.2583E+04	0.0000E+00	2.2736E+04
22	L-STS	-3.3455E+04	-1.7764E+04	-6.8983E+04	0.0000E+00	0.0000E+00	0.0000E+00	-2.6096E+04	-7.6341E+04	0.0000E+00	6.7206E+04
		0.0000E+00	0.0000E+00	0.0000E+00	-1300.	-3139.	0.0000E+00	3397.	-3397.	0.0000E+00	5884.
		3.3455E+04	1.7764E+04	6.8983E+04	0.0000E+00	0.0000E+00	0.0000E+00	7.6341E+04	2.6096E+04	0.0000E+00	6.7206E+04
23	L-STS	7.8000E+04	-3164.	1.3826E+04	0.0000E+00	0.0000E+00	0.0000E+00	7.8155E+04	1.3670E+04	0.0000E+00	7.2296E+04
		0.0000E+00	0.0000E+00	0.0000E+00	-98.95	-69.23	0.0000E+00	120.8	-120.8	0.0000E+00	209.2
		-7.8000E+04	3164.	-1.3826E+04	0.0000E+00	0.0000E+00	0.0000E+00	-1.3670E+04	-7.8155E+04	0.0000E+00	7.2296E+04
24	L-STS	3.8796E+04	-649.8	2.5047E+04	0.0000E+00	0.0000E+00	0.0000E+00	3.8826E+04	2.5017E+04	0.0000E+00	3.4088E+04
		0.0000E+00	0.0000E+00	0.0000E+00	-51.90	-1409.	0.0000E+00	1410.	-1410.	0.0000E+00	2441.
		-3.8796E+04	649.8	-2.5047E+04	0.0000E+00	0.0000E+00	0.0000E+00	12.5017E+04	-3.8826E+04	0.0000E+00	3.4088E+04
25	L-STS	1.6493E+04	-1526.	-1.7571E+04	0.0000E+00	0.0000E+00	0.0000E+00	1.6551E+04	-1.7640E+04	0.0000E+00	2.9623E+04
		0.0000E+00	0.0000E+00	0.0000E+00	114.7	-2559.	0.0000E+00	2561.	-2561.	0.0000E+00	4436.
		-1.6493E+04	1526.	1.7571E+04	0.0000E+00	0.0000E+00	0.0000E+00	1.7640E+04	-1.6561E+04	0.0000E+00	2.9623E+04
26	L-STS	-1.8664E+04	-2973.	-8.3773E+04	0.0000E+00	0.0000E+00	0.0000E+00	-1.8528E+04	-8.3909E+04	0.0000E+00	7.6350E+04
		0.0000E+00	0.0000E+00	0.0000E+00	-466.3	-3484.	0.0000E+00	3515.	-3515.	0.0000E+00	6088.
		1.8664E+04	2973.	8.3773E+04	0.0000E+00	0.0000E+00	0.0000E+00	8.3909E+04	1.8528E+04	0.0000E+00	7.6350E+04
27	L-STS	5.7064E+04	-1693.	2.0329E+04	0.0000E+00	0.0000E+00	0.0000E+00	5.7142E+04	2.0251E+04	0.0000E+00	5.0181E+04
		0.0000E+00	0.0000E+00	0.0000E+00	34.47	-774.3	0.0000E+00	775.1	-775.1	0.0000E+00	1343.
		-5.7064E+04	1693.	-2.0329E+04	0.0000E+00	0.0000E+00	0.0000E+00	-2.0251E+04	-5.7142E+04	0.0000E+00	5.0181E+04
28	L-STS	2.8888E+04	-1190.	2660.	0.0000E+00	0.0000E+00	0.0000E+00	2.8942E+04	2606.	0.0000E+00	2.7731E+04
		0.0000E+00	0.0000E+00	0.0000E+00	-59.31	-1954.	0.0000E+00	1955.	-1955.	0.0000E+00	3386.
		-2.8888E+04	1190.	-2660.	0.0000E+00	0.0000E+00	0.0000E+00	-2606.	-2.8942E+04	0.0000E+00	2.7731E+04
29	L-STS	-996.3	-2362.	5.0486E+04	0.0000E+00	0.0000E+00	0.0000E+00	-883.8	-5.0599E+04	0.0000E+00	5.0163E+04
		0.0000E+00	0.0000E+00	0.0000E+00	-195.0	-3016.	0.0000E+00	3022.	-3022.	0.0000E+00	5234.
		996.3	2362.	-5.0486E+04	0.0000E+00	0.0000E+00	0.0000E+00	5.0599E+04	883.8	0.0000E+00	5.0163E+04

STATIC ANALYSIS EXAMPLE - PRESSURE VESSEL LID

LOADING CASE 1 (SEQUENCE 1) INTERNAL PRESSURE LOADING

AVERAGED PRINCIPAL AND VON MISES STRESSES AT NODES OF PLATE AND SHELL ELEMENTS ORDERED
ACCORDING TO MAXIMUM PRINCIPAL STRESS

NODE	***** PRINCIPAL STRESSES *****		VON MISES
4	83909.	-83909.	0.00000E+00	76350.
26	83909.	-83909.	0.00000E+00	76350.
15	83773.	-83773.	0.00000E+00	76176.
1	78155.	-78155.	0.00000E+00	72296.
23	78155.	-78155.	0.00000E+00	72296.
12	78000.	-78000.	0.00000E+00	72088.
11	76341.	-76341.	0.00000E+00	67206.
22	76341.	-76341.	0.00000E+00	67206.
8	70839.	-70839.	0.00000E+00	63023.
19	70839.	-70839.	0.00000E+00	63023.

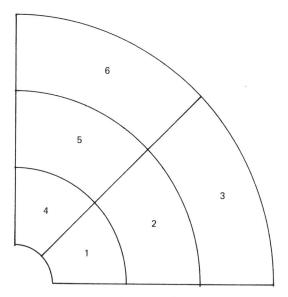

FIGURE 4.55 Static analysis example: pressure vessel lid, plot 3.

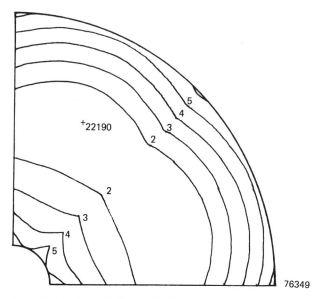

FIGURE 4.56 Static analysis example: pressure vessel lid, superb post plot 4, loading case 1, von Mises equation stress.

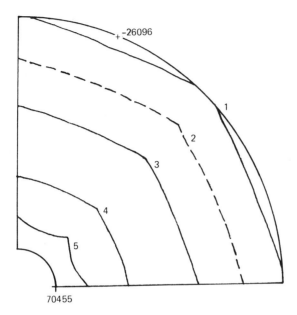

FIGURE 4.57 Static analysis example: pressure vessel lid, superb post plot 5, loading case 1, maximum principal stress.

with their respective numbers. The various stress levels based on the von Mises and the maximum principal stress theories are plotted and shown in Figs. 4.56 and 4.57.

Example 2: Wheel Rim

In this example, a steel wheel rim (with a tire pressure of 26 psi) is modeled using a graphics model from the graphics file which was

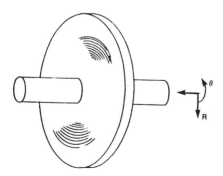

FIGURE 4.58

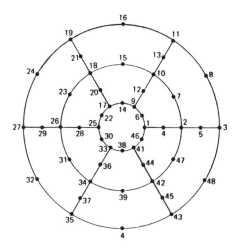

FIGURE 4.59

created on an interactive computer graphics system (Figs. 4.58-4.59). The wheel finite element model is prepared for the NASTRAN analysis program. After the analysis is performed, the results are returned to the output display (display screen) for deformed geometry and stress contour displays (Figs. 4.60-4.70).

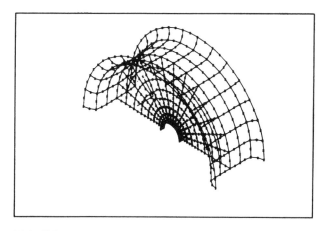

FIGURE 4.60

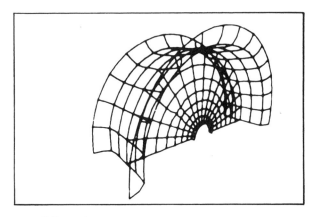

FIGURE 4.61

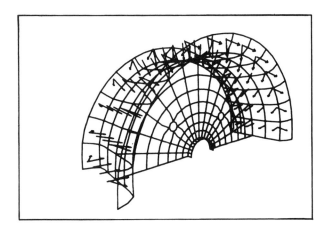

FIGURE 4.62

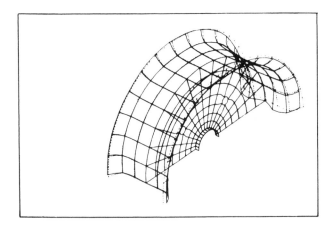

FIGURE 4.63

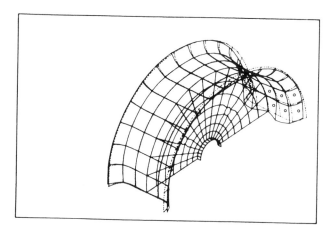

FIGURE 4.64

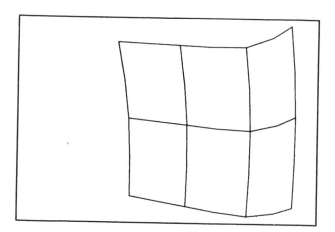

FIGURE 4.65

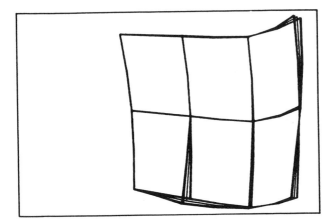

FIGURE 4.66

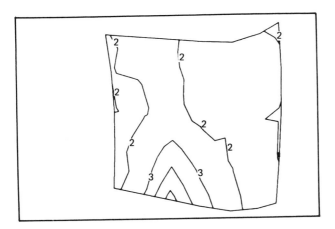

FIGURE 4.67

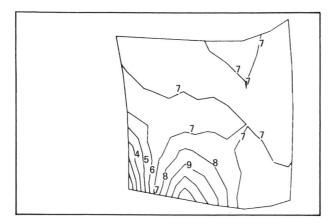

FIGURE 4.68

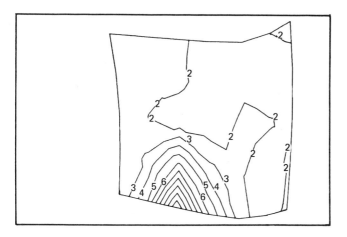

FIGURE 4.69

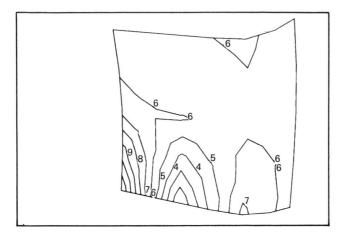

FIGURE 4.70

5
Design of Mechanisms and Machinery

5.1 INTRODUCTION

Physics deals with such topics as heat, light, sound, electricity, magnetism, and mechanics. The topic of mechanics can be divided into two divisions called statics and dynamics. Furthermore, the division of dynamics branches into kinematics and kinetics.

The science of *mechanics* deals with the motion of bodies and forces, and the effect of these forces on the bodies on which they act. *Statics* is that division which deals with forces and with the effects of forces acting on rigid bodies at rest (without motion). *Dynamics* deals with motion and with the effects of forces acting on rigid bodies in motion.

The branch of *kinematics* is the study of motion without consideration of the forces causing the motion and therefore is only concerned with position, displacement, velocity, and acceleration of the body. *Kinetics* is the study of forces acting on rigid bodies in motion and their effect in changing such motion.

5.2 DISPLACEMENT OF MECHANISMS

First, the displacement of the links that make up the mechanism must be analyzed throughout the mechanism's working cycle. This analysis determines the actual path of the various links and how they move relative to each other.

To simplify this analysis, the various links of the mechanism are drawn using single lines to form the configurations of each particular link. Figure 5.1 shows an example of a complex mechanism and its simplified kinematic representation.

(a)

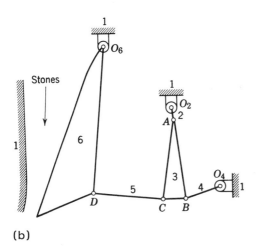

(b)

FIGURE 5.1 (a) Jaw crusher mechanism (courtesy of Allis-Chalmers Manufacturing Co., West Allis, Wis.); (b) kinematic representation of jaw crusher mechanism.

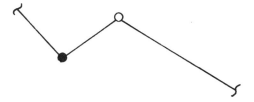

FIGURE 5.2

There are common graphic conventions associated with mechanisms. These graphic conventions are as follows:

A pin joint is represented as a small circle or a dark point (Fig. 5.2).

A simple link is shown as a straight line drawn between the two joints (Fig. 5.3).

A complex link is represented as a figure bounded with straight lines with its pin joints. The figure is usually crosshatched (Fig. 5.4 and 5.5).

Stiff links are links that extend past a joint and are represented as Figs. 5.6 and 5.7.

Fixed or ground joints are pivots attached to the frame or fixed link with ground marks representing the frame (Fig. 5.8).

Sliders are links that slide along another link as shown in Fig. 5.9.

Kinematic drawing consists of determining the complete motion of a given mechanism throughout its entire cycle. This is accomplished by moving and copying each link of the mechanism at each interval of its cycle, as shown in Fig. 5.10. A displacement diagram or time-displacement plot can be made of some point or link in a mechanism with respect to the driver (input link). The time usually consists of one

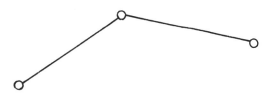

FIGURE 5.3

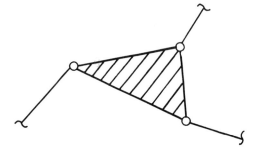

FIGURE 5.4

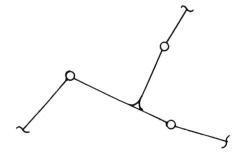

FIGURE 5.5

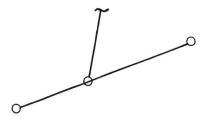

FIGURE 5.6

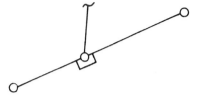

FIGURE 5.7

FIGURE 5.8

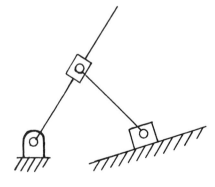

FIGURE 5.9

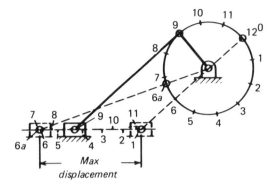

FIGURE 5.10

revolution of the driver laid out horizontally along the y axis. The
displacement units may be linear or angular, depending on the particular motion of the follower being analyzed. Figure 5.11 shows the
time-displacement diagram for the mechanism shown in Fig. 5.10.

It is advantageous to have a displacement curve available for a particular mechanism for several reasons: (1) it provides a graphic picture of the motion of a particular point or link during a complete
cycle of the mechanism, and (2) it is possible to examine this curve
and estimate where the peak velocities and accelerations occur since
the velocity is proportional to the slope and the acceleration is inverse-
ly proportional to the radius of curvature.

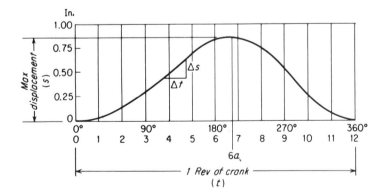

FIGURE 5.11 Displacement diagram: linear displacement. $V \approx \Delta S / \Delta t$.

5.3 DISPLACEMENT, VELOCITY, AND ACCELERATION

The *displacement* of a body is its change of position with reference to a fixed point. Both direction and distance are necessarily stated in order to define completely the displacement of a point or body.

Velocity is the rate of change of position or displacement of a body. A body may change its position by translation through space or by angular movement. Thus it may have linear or angular velocity.

Linear velocity is the rate of linear displacement of a point or body along its path of motion. It consists of speed and direction of motion. A linear velocity can be represented graphically by a vector, the direction showing the direction of movement and the length representing the magnitude of the velocity.

Angular velocity is the rate of change of angular position of a body or line. The sense of rotation is either clockwise (CW) or counterclockwise (CCW). Angular velocity is commonly measured by the angle turned through per unit time. The angle may be expressed in radians, degrees, revolutions, or cycles.

The linear velocity v of a point on a moving body is equal to the angular velocity ωof the body multiplied by the distance of the point from the center of rotation r, or

$$v = \omega r$$

Acceleration is the rate of change of velocity with respect to time. Since velocity may be either linear or angular, also, acceleration can be either linear or angular.

Linear acceleration is the rate of change of linear velocity. Linear acceleration, like linear velocity, can be represented graphically by a vector which shows the magnitude and direction.

Angular acceleration is the rate of change of the angular velocity. It is referred to an angular acceleration if it is in the same direction as the angular velocity and as *deceleration* if it is in the opposite direction of the angular velocity.

For linear motion, the relationships among displacement s, velocity v, and acceleration a may be expressed mathematically as follows:

$$v = \frac{ds}{dt} \qquad a = \frac{dv}{dt} = \frac{d^2 s}{dt^2}$$

Here v is the instantaneous velocity, or velocity at a certain instant, and ds/dt expresses the rate of change of the displacement. Similarly, a is the instantaneous acceleration and dv/dt, or $d^2 s/dt^2$, is the corresponding rate of change of velocity.

When angular motion is dealt with, the displacement is θ, the velocity ω, and the acceleration α. Then, by definition,

$$\omega = \frac{d\theta}{dt} \quad \alpha = \frac{d\omega}{dt} = \frac{d^2\theta}{dt^2}$$

When the motion of a particle or body starts with an initial velocity v_0 or ω_0 and is uniformly accelerated to a velocity v or ω in time t:

$$s = v_0 t + \frac{1}{2} t^2 \quad \text{or} \quad \theta = \omega_0 t + \frac{1}{2} \alpha t^2$$

$$v = v_0 + at \quad \text{or} \quad \omega = \omega_0 + \alpha t$$

$$v^2 = v_0^2 + 2as \quad \text{or} \quad \omega^2 = \omega_0^2 + 2\alpha\theta$$

If the initial velocity, v_0 or ω_0, is zero, that is, the body or particle starts from rest, these equations become

$$s = \frac{1}{2} t^2 \quad \text{or} \quad \theta = \frac{1}{2} \alpha t^2$$

$$v = at \quad \text{or} \quad \omega = \alpha t$$

$$v^2 = 2as \quad \text{or} \quad \omega^2 = 2\alpha\theta$$

The velocity of a moving point may change in two ways: (1) its linear speed along its path may increase or decrease, or (2) the direction of its motion may change.

1. The rate of change of speed in the direction of motion is the tangential acceleration, since this involves an acceleration acting along the path of motion.
2. The change in the direction of the motion is due to the normal or centripetal acceleration, which acts in a direction normal to the direction of the path of motion. These terms may therefore be defined as follows:
   a. The *tangential acceleration* is the linear acceleration in the direction of motion at the instant considered and is measured by the rate of change of speed along its path.
   b. The *normal* or *centripetal acceleration* is that acceleration which causes the direction of motion of a body to change. It acts

along a line perpendicular to the path of motion and toward the center of curvature of this path. The normal acceleration equals the change in normal velocity/dt, where $a^n = v\omega dt/dt = v\omega = v^2/r$ or $r\omega^2$.

c. The tangential acceleration of a point on a moving body is equal to the angular acceleration of the body multiplied by the distance from the point to the center of rotation.

5.4 INSTANT CENTERS

Links of machines with plane motion may be divided into three groups: (1) those with angular movement but not about a fixed axis, (2) those with angular movement but not about a fixed axis, and (3) those with linear but not angular motion. All these motions may be studied by the use of instant centers. For kinematic purposes, the thickness of the bodies perpendicular to the plane of motion will be disregarded and we will deal with the projections of the bodies on this plane.

The *instant center* may be defined in either of the following ways: (1) when two bodies have plane relative motion, the instant center is a point on one body which the other rotates at the instant considered; and (2) when two bodies have plane relative motion, the instant center is the point at which the bodies are relatively at rest at the instant considered.

These centers may easily be located by means of *Kennedy's theorem*. This theorem states that the instant centers for any three bodies having plane motion lie along the same straight line. In any mechanism having plane motion, there is one instant center for each pair of links. The number of instant centers is equal to the number of pairs of links. With n links, the number of instant centers is equal to the number of combinations of n objects taken two at a time, namely $n(n - 1)/2$.

For example, when a mechanism has six links, the number of instant centers to be located is 15. It is then desirable to have a systematic method of recording progress and assisting in the determination. This may be done by means of a circle diagram or by the use of tables. Both methods will be given and illustrated by an example. A diagram of the form shown in Fig. 5.11 is useful when finding instant centers since it gives a visual indication of the order in which the centers can be located by means of Kennedy's theorem and also, at any stage in the process, it shows what centers remain to be found. The circle diagram will be used for finding the centers in the six-link mechanism of Fig. 5.12a. The following procedure is used to locate them.

Draw a circle, as in Fig. 5.12c, and mark points, 1, 2, 3, 4, 5, and 6 around the circumference, representing the six links in the mechanism. As the instant centers are located, draw lines connecting the points with corresponding numbers on this diagram. Thus line 12

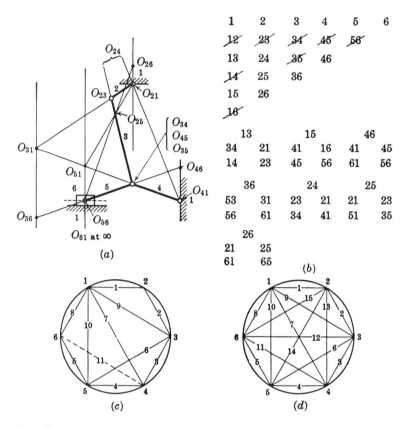

FIGURE 5.12 Crusher mechanism.

is drawn when the instant center O_{12} is found. The figure will have lines connecting all pairs of points when all the instant centers have been determined. Numbers on the lines, indicating the sequence in which they are drawn, facilitate checking. At one stage in the process (after 10 of the centers have been found) the diagram will appear as shown in Fig. 5.12c. From an inspection of the diagram, it is noted that by joining 46, two triangles, 465 and 461, are completed. Since this is the case, then, locate the instant center O_{46} at the intersection of $O_{41}O_{61}$ and $O_{45}O_{56}$. If instead, 62 was drawn, only one triangle, namely 621, would have been formed; therefore, the center O_{62} could not be found at this stage, although it can be after O_{25} (line 14) is placed. Therefore, the line 62 is taken as number 15. The procedure is the same for the remaining points.

If each line is first drawn dashed while the center is being located and made solid as soon as the center is found, errors are less likely to occur. Figure 5.12a shows the location of all the instant centers and Fig. 5.12d the complete circle diagram.

An alternative method of locating instant centers that is in common use is the *tabular method*. In this procedure a main table is set up and used with supplementary tables, as illustrated in Fig. 5.12b. Across the top of the main table are listed the numbers of the links in the mechanism. In the first column are listed the number at the top of the column combined with those numbers to the right of it. In the second column the number at the top of that column combined with those to the right. Continuing this procedure to the end of the table gives a complete list of all the centers to be found. As the centers are located on the drawing, they are crossed off on the table, as shown. Usually, about haf the centers may be found by inspection and crossed off immediately. Thus, in the example of Fig. 5.12, eight of the centers—O_{12}, O_{23}, O_{34}, O_{45}, O_{56}, O_{14}, O_{16}, and O_{35}—may be found by inspection. The remainder will have to be found by the use of Kennedy's theorem and with the aid of supplementary tables. Suppose now that it is desired tolocate the center O_{31}. A supplementary table is set up in which links 1 and 3 are considered with a third link, say 4. Then the centers O_{34}, O_{14}, and O_{13} must lie on a straight line, by Kennedy's theorem. The third link might also be 2, when the centers O_{21}, O_{23}, and O_{13} will lie on a straight line. These centers are listed in a supplementary table under the heading 13. Reference to the main table shows that the centers O_{34}, O_{14}, O_{21}, and O_{23} have been crossed off—thus have been located and are available. Drawing lines through them locates O_{31}. In a similar manner, by using the tables, all of the centers may be located. The supplementary tables in Fig. 5.12d show the procedure.

Frequently, it may be found that the third link chosen may require centers that have been located. In such cases another third link would have to be tried. In the early stages it may be found that no third link will be satisfactory. In that case the search for that particular center will have to be abandoned temporarily until more centers are formed.

5.5 VELOCITY AND ACCELERATION IN PLANE MOTION

To illustrate the various methods of finding velocities and accelerations, consider the six-link mechanism shown in Fig. 5.13a, on which most of the instant centers are lcoated. The velocity of point R is given by the vector v_R, and it is desired to find the velocity of point T by (1) the connecting-link method; (2) the direct method (considering R as a point on link 2, and T as a point on link 6); (3) the resolution

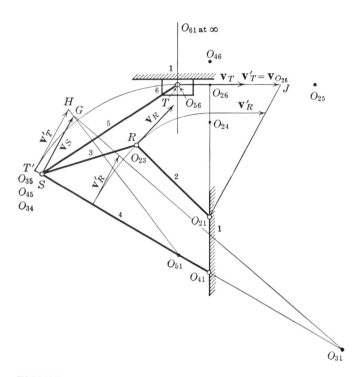

FIGURE 5.13a

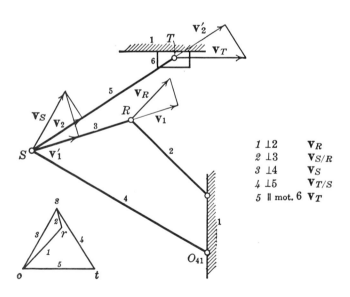

$1 \perp 2 \quad \mathbf{v}_R$
$2 \perp 3 \quad \mathbf{v}_{S/R}$
$3 \perp 4 \quad \mathbf{v}_S$
$4 \perp 5 \quad \mathbf{v}_{T/S}$
$5 \parallel \text{mot.} \; 6 \; \mathbf{v}_T$

FIGURE 5.13b

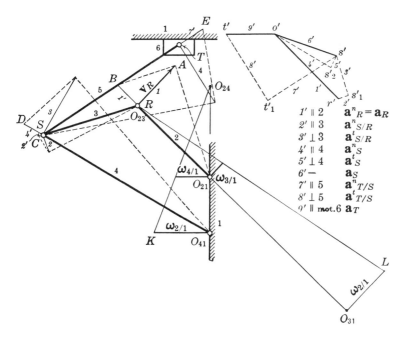

FIGURE 5.13c

method; and (4) the velocity-image method, and to find the acceleration of point T by (5) the acceleration-image method.

Connecting-Link Method

Points R and S are both on link 3 and therefore rotate about the
pivot center O_{31}. On Fig. 5.13a, point R is swung about O_{31} to link
4, and the triangle $O_{31}GS$ drawn to find the velocity of S. Both S and
T are points on link 5 and therfore rotate about the pivot center O_{51}.
Point T is swung about O_{51} to link 4 extended, and the triangle $O_{51}HT'$
is drawn to find the velocity v'_T. This vector is then swung back
about O_{51} to point T to give the velocity of T.

Direct Method

Considering R as a point on link 2 and T as a point on link 6, the
three instant centers to be used are O_{21}, O_{26}, and O_{61}. These are
located on Fig. 5.13a. Points R and the instant center O_{26} are both
points on link 2 and therefore rotate about the pivot center O_{21}. The
vector v_R is then swung about O_{21} to a line through $O_{21}O_{26}$, and the
triangle $O_{21}O_{26}$ is drawn to find the velocity of O_{26}. This should then
be related in a similar manner to point T. However, the pivot center

O_{61} is at infinity, which means that any point on link 6 has the same magnitude and direction of velocity. Therefore, the velocity of O_{26}, a point on link 6, is the same as the velocity of T, and the vector is merely transferred to point T.

Resolution Method

The velocity of v_R of point R may be broken into two components: one along link 3, which is labeled v_1, and one perpendicular to this link, as shown in Fig. 5.13b. The component along the link must be the same at either end and therefore is shown as v_1' at point S. From the end of the v_1' component at S, a normal line is drawn to link 3 representing the other component. Where it intersects the resultant velocity vector of S, which is perpendicular to link 4, determines the velocity of S. Again this resultant velocity of S may be broken into two components, one along link 5, which is labeled v_2, and the other perpendicular to link 5. The component along the link must be the same at either end and therefore is also shown at point T as v_2'. The absolute velocity of point T must be horizontal, as that is the only direction in which it can move. Drawing a normal at the end of the v_2' vector at its tip determines the magnitude of the velocity of point T, or v_T.

Velocity-Image or Relative-Velocity Method

Starting from a pole o (Fig. 5.13b), the given velocity of point R is laid off perpendicular to link 2. The velocity of point S relative to R is in a direction perpendicular to line RS, or link 3. Therefore, from r on the diagram draw line 2 perpendicualr to link 3. The velocity of point S is in a direction perpendicular to link 4. As the velocity of S is an absolute one, or relative to the fixed link, line 3 is drawn from pole o perpendicular to link 4. The intersection of lines 2 and 3 locates points s; and os, or line 3, is the velocity of point S.

The velocity of point T relative to S is in a direction perpendicular to link 5; therefore, line 4 is drawn from s normal to link 5. The velocity of point T must be horizontal. Since it is relative to the fixed link, it is drawn from the pole o. The intersection of lines 4 and 5 determines point t; and ot, or line 5, represents the velocity of point T.

Acceleration-Image Method

The absolute and relative velocities needed in drawing this diagram are taken from the velocity-image diagram of Fig. 5.13b. As the velocity of point R is assumed to be constant, it has only a normal acceleration. Its magnitude is found by the right-angled triangle ABO_{21} to be BR (see Fig. 5.13c). The acceleration image is drawn triple size for clarity; therefore, this distance is laid off from the pole o' parallel to link 2 triple size to give line 1', or o'r'.

Point S will have both a normal and tangential acceleration relative to point R. The normal acceleration is found by laying off line 2 of the velocity image at S; and the distance CS, found by the right-triangle construction, is laid off triple size from r' parallel to link 3 to give line 2'. The tangential acceleration of S relative to R is perpendicular to this line and is labeled 3'. The absolute acceleration of point S relative to the ground is the vector sum of the normal and tangential accelerations. If line 3 is used of the velocity image at point S, the length SD is the normal acceleration of S relative to the fixed link 1. Since this is an absolute value, it is laid off triple size from the pole o' parallel to link 4 as line 4'. The tangential acceleration of S relative to the ground is normal to link 4 and is drawn as line 5'. The intersection of lines 3' and 5' locates s'. Line 6' from the pole o' to s' represents the absolute acceleration of point S.

Point T has both a normal and tangential acceleration relative to point S. Placing line 4 of the velocity image at point T and drawing the right-angled triangle gives TE as the normal acceleration of T relative to S. This is laid off triple size as line 7' from s' parallel to link 5. The tangential acceleration is perpendicular to this and is shown on line 8'. Any absolute acceleration of the slider 6, or point T, must be horizontal. Therefore, line 9', or o't', is then the absolute acceleration of point T. The tabulation given in Fig. 5.15c summarizes the steps just outlined.

5.6 CORIOLIS ACCELERATION

The image method of determining acceleration is applicable only to points located on a rigid body. Referring to the mechanism shown in Figs. 5.14a and 5.14b, the following vector equation can be written:

$$\underline{a}_{Q/1} = \underline{a}_{P/1} \rightarrow \underline{a}_{Q/P} = \underline{a}^n_{P/1} \rightarrow \underline{a}^t_{P/1} \rightarrow \underline{a}^n_{Q/P} \rightarrow \underline{a}^t_{Q/P}$$

In writing this equation, which is the basis of the acceleration of this image, the points located on the rigid link 3 are being used.

Occasionally, problems arise in which it is necessary to find the acceleration of points not on the same rigid body. For such problems it is necessary to use the Coriolis law. This is illustrated in Fig. 5.15a, where there is a cam rotating at a constant angular velocity of 20 rad/sec counterclockwise. A flat face on the cam is in contact with a roller follower and it is desired to determine the acceleration of the follower for the position shown.

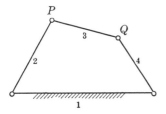

FIGURE 5.14a

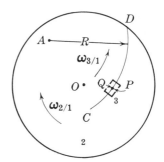

FIGURE 5.14b

To obtain the coincident points needed in the analysis, it is imagined the cam (link 2) is extended to include the center of the roller B of the follower (link 3). Therefore, coincident points B2 and B3 are obtained on links 2 and 3, respectively. The path of point B3 on link 2 will be a straight line parallel to the flat face of the cam; that is, considering the cam fixed and the follower rolling along the flat face, its center B3 will trace a straight line parallel to the cam face as shown by the dashed line.

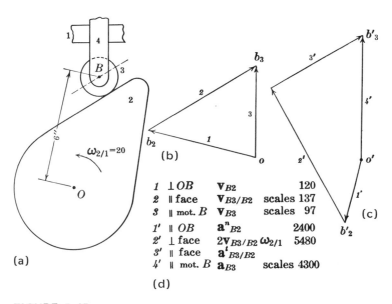

1	$\perp OB$	$\mathbf{V}_{B2}$		120
2	‖ face	$\mathbf{V}_{B3/B2}$	scales	137
3	‖ mot. B	$\mathbf{V}_{B3}$	scales	97
1'	‖ OB	$\mathbf{a}^n_{B2}$		2400
2'	$\perp$ face	$2\mathbf{V}_{B3/B2}\,\omega_{2/1}$		5480
3'	‖ face	$\mathbf{a}'_{B3/B2}$		
4'	‖ mot. B	$\mathbf{a}_{B3}$	scales	4300

(d)

FIGURE 5.15

The velocity triangle may now be drawn as shown in Fig. 5.15b. The velocity of point B2 equals $OBW_{2/1} = 6 \times 20 = 120$ in./sec; the direction of the velocity of B3 is known to be vertically upward, whereas that of $v_{B3/B2}$ is parallel to the cam face. Lines 2, or $v_{B3/B2}$, and 3, are scaled as 137 and 97 in./sec, respectively.

Using the Coriolis expression, the following equation can be written:

$$\underline{a}_{B3/1} = \underline{a}_{B2/1} \overset{\rightarrow}{+} \underline{a}_{B3/B2} = \underline{a}^n_{B2/1} \overset{\rightarrow}{+} \underline{a}^t_{B2/1} \overset{\rightarrow}{+} \underline{a}^n_{B3/B2}$$

$$\overset{\rightarrow}{+} 2\underline{v}_{B3/B2}W_{2/1}$$

It is calculated that

$$\underline{a}^n_{B2/1} = OBW_{2/1} = 6(20)^2 = 2400 \text{ in./sec}^2$$

and it acts downward in a direction parallel to OB and $a^t_{B2/1} = 0$, since $W_{2/1}$ is constant while $a^n_{B3/B2} = 0$, as the direction of $V_{B3/B2}$ is along a straight line, or the radius of this rotation is infinite. If the cam surface were curved at the point of contact with the follower, it is necessary to determine its radius of curvature R and use the equation $a^n_{B3/B2} = v^2_{B3/B2}/R$. The magnitude of $a^t_{B3/B2}$ is unknown, but its direction is parallel to the cam face, or path of motion of B_3 relative to link 2.

The Coriolis component

$$2\underline{v}_{B3/B2}W_{2/1} = 2 \times 137 \times 20 = 5480 \text{ in./sec.}^2$$

its direction being perpendicular to the cam surface and acting upward to the left (i.e., in the direction of the $\underline{v}_{B3/B2}$ vector rotated 90° in the direction of $w_{2/1}$). The acceleration of point B_3 must be vertically upward; therefore, the diagram may be drawn as shown in Fig. 5.15c. Point b'_3 lies at the intersection of lines 3' and 4'. The supplementary table (Fig. 5.15d) outlines the procedure. The distance $o'b'_3$ represents the acceleration of point B_3 and the follower. The distance scales as 4300 in./sec².

5.7 FORCE ANALYSIS OF MECHANISMS

An important part of mechanical design is determining the strength of the various parts of a mechanism. Therefore, it is necessary to determine the forces and torques acting on the individual links of the mechanism by the use of the free body with its associated equilibrium equa-

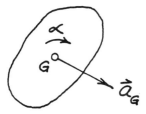

FIGURE 5.16

tions. This analysis must be performed on each component of a complete mechanism; that is, in reality, a four-link mechanism consists of eight links if the pins or bearings connecting the primary members are included. Bearings, pins, screws, and so on, are often critical elements in mechanisms because of the concentration of forces at these elements.

One must realize that mechanisms operating at high speeds often produce inertia forces on individual links which are greater than the static forces that are present. These inertia forces, which are expressed in terms of the accelerated motion of the individual links, must therefore be taken into account in the force analysis. This type of analysis is commonly known as *kinetics*.

Figure 5.16 shows a body in general plane motion, whose state of acceleration, characterized by a_G and α, may be attributed to the action of some two-dimensional force system, as shown in Fig. 5.17.

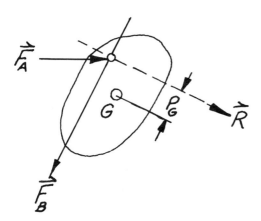

FIGURE 5.17

The magnitude, direction, and sense of the R are defined by R = m X a_G and its position with respect to G follows from the torque equation:

$$T_G = d_G R = I_G \alpha = mK_G^2 \alpha$$

with CCW sense positive, yielding

$$d_G = \frac{I_G \alpha}{ma_G} = \frac{K_G^2 \alpha}{a_G}$$

where

m = mass of body
I_G = its moment of inertia about the axis through the body's center of gravity G
K_G = corresponding radius of gyration
d_G = perpendicular distance between G and R

The body may be brought into an apparent state of equilibrium, as shown in Fig. 5.18, by adding tot he actual force system a fictitious force, called the inertia force, which would just balance R. This inertia force is known as *d'Alembert's principle*. Its application makes it possible to analyze the forces acting on a moving body, or a system of connected bodies, by the more easily handled principles of statics.

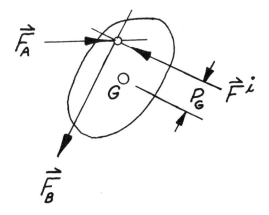

FIGURE 5.18

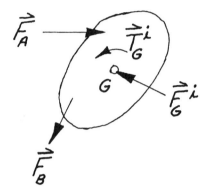

FIGURE 5.19a

The magnitude, direction, and sense of the inertia force are given by $F^i = -m \times a_G$ and its position with respect to G is defined by equation $d_G = I_G \alpha/m \; a_G = K_G^2 \alpha/a_G$ and the fact that the sense of the torque of F^i is opposite to that of α.

As an alternative, the equilibrant inertia force F^i may be replaced by a force F_G^i through the center of gravity, an inertia couple of torque T_G^i, as shown in Figs. 5.19a and 5.19b, where

$$F_G^i = -m \times a_G \quad \text{and} \quad T_G^i = -I_G \times \alpha$$

In the case of a body rotating about a fixed axis, shown in Fig. 5.20, the inertia force system consists, as before, of a single force

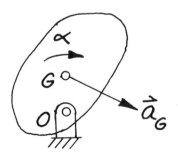

FIGURE 5.19b

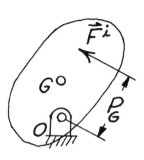

FIGURE 5.20

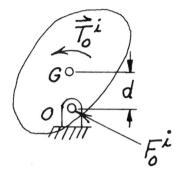

FIGURE 5.21

$F^i = -m \times a_G$ whose position with respect to point O follows from the torque equation $d_0 F^i = -I_0 \alpha = -mK^2 \alpha$, which yields $d_0 = I_0 \alpha / m \times a_G = K^2/a_G$ where

I_0 = moment of inertia of body about the axis through O, lb-in.-sec^2

K_0 = corresponding radius of gyration, in.

Alternatively, as shown in Fig. 5.21, F^i may be replaced by the force F_0^1 through O and a couple of torque T_0^i, where $F_0^i = -ma_G$ and $T_0^i = -I_0 \alpha$.

If a system of connected bodies, such as a mechanism, is in equilibrium, then each individual link is also in equilibrium. Therefore, in accordance with d'Alembert's principle, each link may be considered to be in equilibrium under the combined action of the active forces directly applied to it, its own inertia force, and the reactive forces (usually unknown in the beginning) exerted on it by the paired links. Provided that all these forces are introduced, the link may be regarded as a free body (i.e., disconnected from the rest of the mechanism). Such free-body diagrams are an essential tool in the force analysis of mechanisms. For each link, three equilibrium equations may be established, and if the particular link is statically determinate, the unknown force components can be obtained either analytically or graphically. If the link, taken by itself, is statically indeterminate, it must be considered in conjunction with one or more of its paired neighbors.

In this connection it should be noted that lower kinematic pairs are inherently associated with two unknown quantities: the magnitude of the transmitted force and its diregion (turning and rolling pairs) or its position (sliding pairs), whereas in higher (skidding) pairs the position and direction are easily obtained and only the magnitude is unknown.

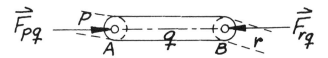

FIGURE 5.22

Figure 5.22 shows a link q, the only forces acting on which are those exerted by the links p and r, paired with it. Clearly, if q is to remain in equilibrium, then F_{pq} and F_{rq} must be collinear, equal in magnitude, and opposite in sense.

Figure 5.23, a third force, F_q, has been added to the system. The following particulars are now assumed to be known: F_q, line of action, magnitude, sense; F_{pq}, line of action; F_{rq}, point of application. Since only three quantities remain to be established—the magnitude of F_{pq} and the magnitude and direction of F_{rq}—the problem is statically determinate. Application of the torque equation $T = 0$ with respect to the point of intersection of F_q with F_{pq} leads to the conclusion that F_{rq} must also pass through this point. Having thus located the line of action of F_{rq}, the two magnitudes can be obtained either by calculation or by construction (Fig. 5.24).

Figure 5.25a and 5.25b show the system (Fig. 5.23), modified by the addition of a cople of torques T_q and F_q, and T_q may be combined to the single force $F'_q = F_q$, but displaced with respect to F_q by the distance $n = T/F_q$, in a direction determined by the sense of T_q.

The triple-paired, or ternary, link q in Fig. 5.26 is in equilibrium under the action of the completely specified force F_q and the link forces F_{pq}, F_{rq}, and F_{sq}, the lines of action of which are given. As before, the unknown quantities, that is,t he three force magnitudes (and specific directions), can be found by calculation or by construction (Fig. 5.27).

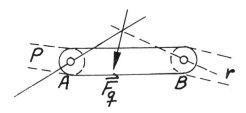

FIGURE 5.23

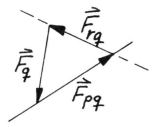

FIGURE 5.24

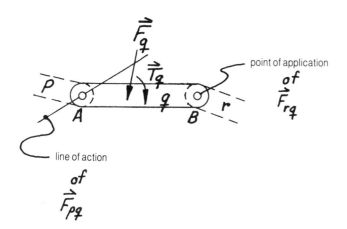

point of application

of

$\vec{F}_{rq}$

line of action

of

$\vec{F}_{pq}$

FIGURE 5.25a

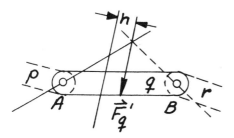

FIGURE 5.25b

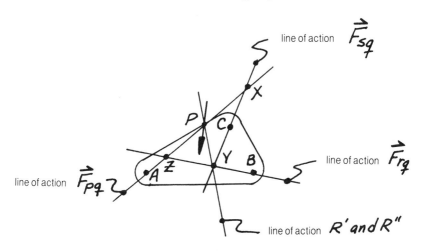

FIGURE 5.26

The following particulars refer to the four-bar mechanism of Figs. 5.28a, 5.28b, and 5.28c:

Link 1: O_2O_4 = 12 in.
Link 2: O_2A = 4 in.; W_2 = 20 lb; G_2 in O_2; W_2 = 40 rad/sec, CCW, constant

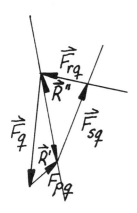

FIGURE 5.27

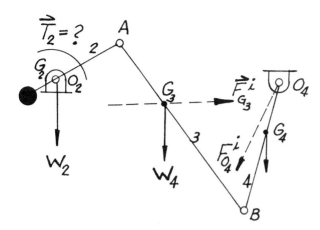

FIGURE 5.28a

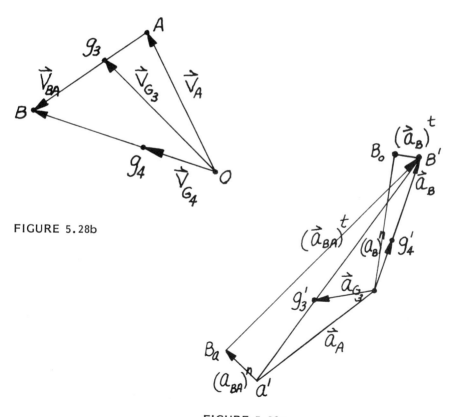

FIGURE 5.28b

FIGURE 5.28c

Link 3: AB = 11 in.; W_3 = 10 lb; k_G = 4.5 in.; AG_3 = 4 in.
Link 4: O_4B = 7 in.; W_4 = 15 lb; k_O = 3.9 in.; O_4G_4 = 2.5 in.;
         T_4 = 200 in.-lb., CCW

Determine all pin forces and the driving torque T_2 at the instant when ϕ_2 = 30°. Solve the problem by considering the force system altogether.

The preliminary step consists of the determination of the inertia effects. The acceleration diagram yields

$$\alpha_3 = 1045 \text{ rad/sec}^2(+) \qquad a_{G_3} = 2620 \text{ in./sec}^2, \text{ IIO'} \rightarrow g_3'$$

$$\alpha_4 = 148.5 \text{ rad/sec}^2(+) \qquad a_{G_4} = 2040 \text{ in./sec}^2, \text{ IIO'} \rightarrow g_4'$$

Therefore,

$$F^i_{G3} = 68 \text{ lb, II } g_3' \rightarrow \text{O'} \qquad T^i_{G3} = -550 \text{ in.-lb}$$

$$F^i_{O4} = 80 \text{ lb, II } g_4' \rightarrow \text{O'} \qquad T^i_{O4} = -88 \text{ in.-lb}$$

Also, from the data,

$$F^i_2 = 0 \quad \text{and} \quad T^i_2 = 0$$

The complete analysis involves the determination of nine quantities: the magnitudes and directions of the four pin forces, and the driving torque. Since nine independent equilibrium equations can be established (three for each moving link), the mechanism, as a whole, is statically determinate. To being the analysis, it is necessary to locate the simplest, in itself statically determinate, group of paired links. Each link, considered as a separate free body, is statically indeterminate (links 3 and 4, four unknown quantities each; link 2, five unknowns). However, it will be recognized that links 3 and 4, taken as a group, form a statically determinate unit (six unknowns, six equations of equilibrium). The solution proceeds as follows:

1. Consider the free-body diagram of link 3 (Fig. 5.29a). The pin force F_{43} can be resolved into the perpendicular components $F^{\perp 3}_{43}$ and $F^{\parallel 3}_{43}$, the first of which can be determined by taking moments about A. (The moment arms of W_4 and F'_{G3} are scaled off the drawing.) In the present case,

$F_{43}^{\perp 3} = 17.4$ lb, directed to the right

2. Consider the free-body diagram of link 4 (Fig. 5.29b). The pin F_{34} can be resolved into the perpendicular components $F_{43}^{\perp 4}$ and $F_{34}^{\parallel 4}$. The former can be determined by taking moments about O_4. Therefore, $F_{34}^{\perp 4} = 4.0$ lb, directed to the left

3. The independent perpendicular components $F_{34}^{\perp 3}$ ($= -F_{43}^{\perp 3}$) and $F_{34}^{\perp 4}$, as shown in Fig. 5.29c. Here $F_{34} = 32.0$ lb, at $221.5°$ to $O_2 \rightarrow O_4$.

4. The equilibrium force polygon for link 4, (Fig. 5.29d) yields $F_{14} = 123$ lb, at $62°$ to $O_2 \rightarrow O_4$.

5. The equilibrium force polygon for link 3 (Fig. 5.29e) gives $F_{23} = 93.0$ lb, at $189.5°$ to $O_2 \rightarrow O_4$.

6. The equilibrium force polygon for link 2 (Fig. 5.29f) and the torque equation $\Sigma\, T_{O2} = 0$ [with the arm p (Fig. 5.29g) scaled off the drawing] yield $F_{12} = 92.0$ lb at $177°$ to $O_2 \rightarrow O_4$ and $T_2 = 131$ in.-lb (+)

The preceding problem can be solved by the method of superposition. The basic principles of the method of superposition will be demonstrated by determining separately the effects of the force systems on each link, rather than by looking for the effects of each individual force and couple.

1. Consider the external forces acting on link 4 alone. For this loading, shown in Fig. 5.30a, link 3 is a two-force memeber and the force F'_{34} that it exerts on link 4 acts along AB. The torque equation $\Sigma\, T_{O4} = 0$ yields $F'_{34} = 21.9$ lb, $\parallel$ B $\rightarrow$ A.

2. The equilibrium force polygon for link 4 (Fig. 5.30b) yields $F'_{14} = 84.5$ lb, at $55°$ to $O_2 \rightarrow O_4$.

3. F'_{34} ($= -F'_{34}$) is transmitted directly through link 3 to link 2, as shown in Fig. 5.30a, so that $F'_{32} = -F'_{34}$. Equilibrium conditions of link 2 require that $F'_{12} = -F'_{32} = F'_{34}$ and $T'_2 = -p'F'_{32} = +87$ in.-lb

4. Next consider the external forces acting on link 3 alone. For this loading, shown in Fig. 5.30c, link 4 is a two-force member. Thus F''_{43} acts along O_4B. The torque equation $\Sigma T_A = 0$ yields $F''_{43} = 40.0$ lb, $\parallel$ B $\rightarrow$ O_4

5. The equilibrium force polygon for link 3 (Fig. 5.30d) shows that $F''_{23} = 84.5$ lb, at $203°$ to $O_2 \rightarrow O_4$.

6. Equilibrium conditions for link 2 require that $F''_{12} = -F''_{32} = F''_{23}$ and $T''_2 = -pF''_{32} = +44$ in.-lb

7. In this step consider the effect of the external forces acting on link 2 alone. The only such force is W_2, which passes through O_2. Therefore, $F'''_{12} = -W_2 = 20$ lb, directed upward.

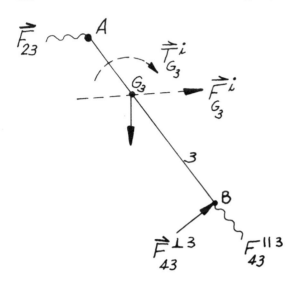

FIGURE 5.29a

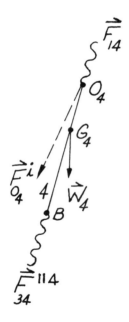

FIGURE 5.29b

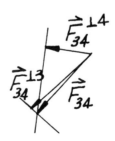

FIGURE 5.29c

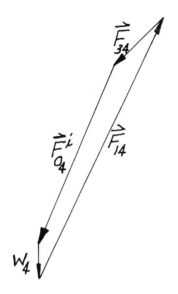

FIGURE 5.29d

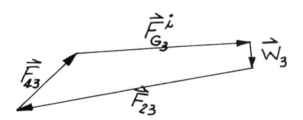

FIGURE 5.29e

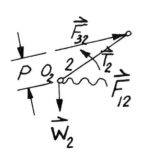

FIGURE 5.29f

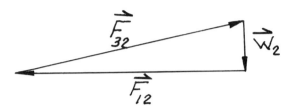

FIGURE 5.29g

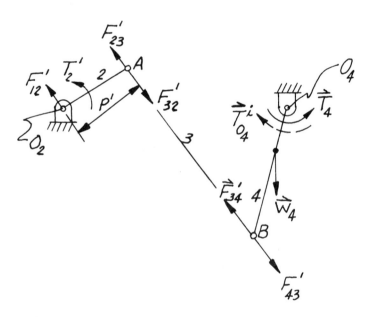

FIGURE 5.30a

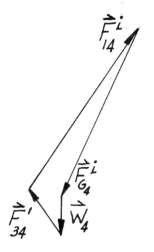

FIGURE 5.30b

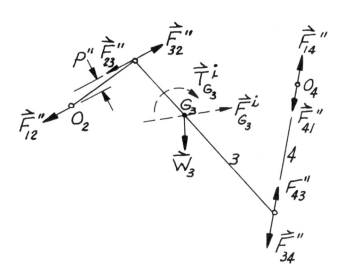

FIGURE 5.30c

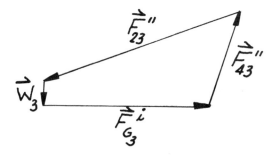

FIGURE 5.30d

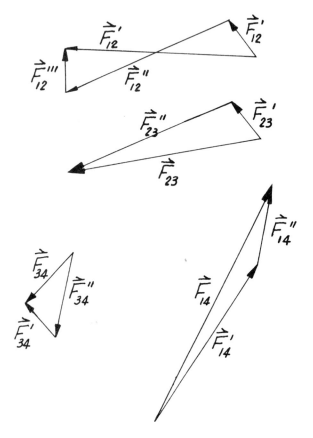

FIGURE 5.31

8. The component effects are added vectorially, as shown in Fig.
    5.31. The results are identical with those of the preceding method.
    Also, $T_2 = T_2' + T_2'' = +131$ in.-lb

5.8 MOTION ANALYSIS USING INTERACTIVE
COMPUTER GRAPHICS

Interactive computer graphics is a useful tool for designing mechanisms.
It is important to understand the various techniques involved when
using interactive computer graphics for kinematic and kinetic analysis
of mechanisms.
    For example, consider the four-bar linkage shown in Fig. 5.32a.
To analyze its entire motion on an interactive computer graphics sys-

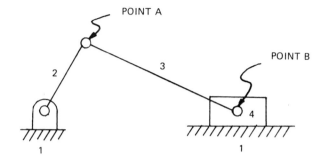

POINT A

POINT B

2

3

4

1

1

FIGURE 5.32a

tem (ICGS), the following procedure would be used. Figure 5.32b shows this four-bar linkage on the display screen of an ICGS and its input link (link 2) being rotated (by 30° increments in a counter-clockwise direction) about its pivot point located on link 1 (ground or frame link). Every time link 2 is rotated, this rotated position is placed on a different layer on the ICGS. In this case, 12 layers

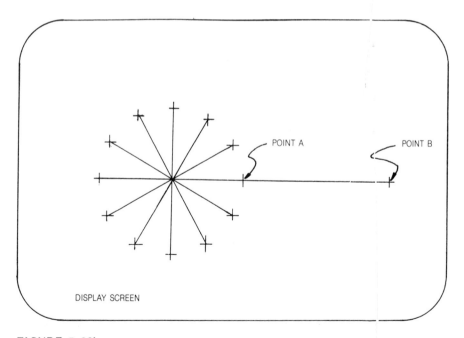

POINT A

POINT B

DISPLAY SCREEN

FIGURE 5.32b

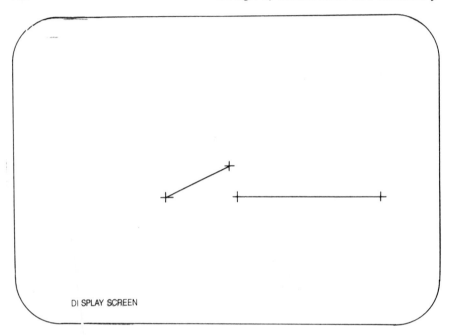

DISPLAY SCREEN

FIGURE 5.33a

(0, 1, 2, 3, 4, 5, 6, 7, 8, 9, 10, and 11) are used to show these rotated positions.

A circle with radii equal to the length of link 3 is placed at the endpoint A of link 2 as shown in Figs. 5.33a through 5.33e. The intersection of this circle with the path of point B on link 4 (represented by
a horizontal line) defines this point B on link 4. This procedure is

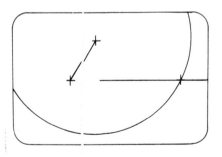
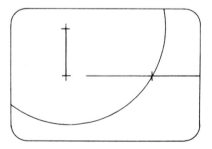

FIGURE 5.33b

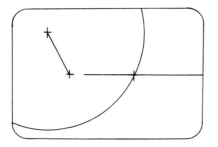
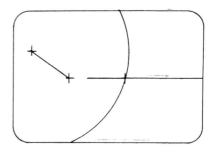

FIGURE 5.33c

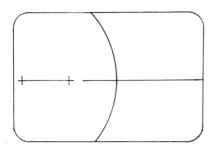
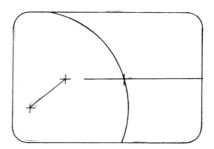

FIGURE 5.33d

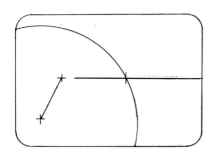
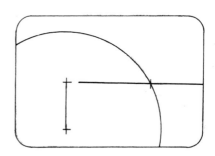

FIGURE 5.33e

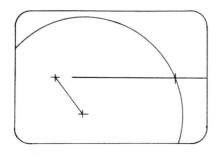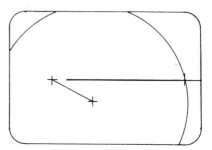

FIGURE 5.34

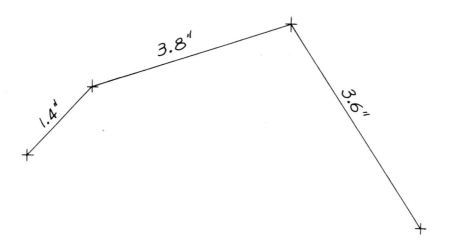

FIGURE 5.35a

LAY 0

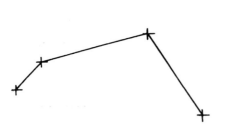

FIGURE 5.35b

LAY 0

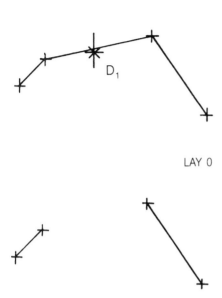

LAY 0

FIGURE 5.35c

repeated for each of the other 10 positions on each of the remaining 10 layers. The results are summarized in Fig. 5.34.

Figures 5.35a through 5.35e show another four-bar linkage. A similar approach is used to obtain the complete motion of this mechanism using an ICGS. For example, the commands used to draw the linkage and analyze its motion was selected from the menu shown in Fig. 5.36. This menu consists of various commands that are acceptable and understood by the particular ICGS for constructing geometry. The location of each command must be digitized in the proper sequence, thus forming an understandable syntax statement for directing the construction of the particular geometry on the ICGS. These commands will be located on the menu for the reader by using a row-column type of matrix notation, enabling the reader to better understand the selection and digitizing process involved on an ICGS.

In addition, BEFORE and AFTER display screens will be used to show the results of each set of commands with the selected command shown next to these display screens. This type of simulation will help the reader understand the actual operation of an ICGS while doing this example.

LAY 1

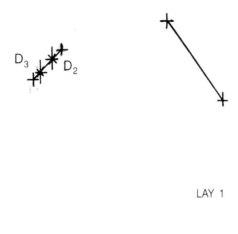

LAY 1

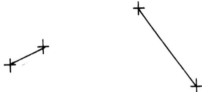

FIGURE 5.35d

The first task is to draw the four-bar linkage shown in Fig. 5.35a
on the ICGS's display screen (Fig. 5.35b). The following commands
are used:

    PICK LAYER 0 [CR]
    PLACE LINE / INPUT X0Y0,X1Y1, X4Y2,X6Y-1 [CR]

Again, link 2, which is the input link, is rotated (in this case, clockwise) 18 increments of 20° per increment. The commands used to revolve link 2 in 20° increments are listed below. One must remember
to place link 2 on a different layer every time it is revolved. Otherwise, there will be too much geometry appearing on the display screen
at the same time, making it almost impossible to read and understand.
Also, link 3 has been erased from the display screen, since it does not
serve any useful purpose in the analysis (Fig. 5.35c).

LAY 2

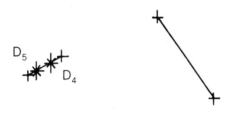

LAY 2

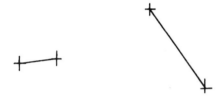

FIGURE 5.35e

ERASE LINE/UNIT D_1 [CR]
COPY LAYER 1 [CR]
PICK LAYER 1 [CR]
REVOLVE ANGLE −20/UNIT D_2 END D_3 [CR]

COPY LAYER 2 [CR]
PICK LAYER 2 [CR]
REVOLVE ANGLE −20/UNIT D_4 END D_5 [CR]

COPY LAYER 3 [CR]
PICK LAYER 3 [CR]
REVOLVE ANGLE −20/UNIT D_6 END D_7 [CR]

etc.

PLACE CIRCLE RADIUS _____/UNIT END D [CR]
PLACE CIRCLE RADIUS/UNIT END D [CR]
PLACE POINT INTERSECTION OF/UNIT DD [CR]
(Figs. 5.35d to 5.35i)

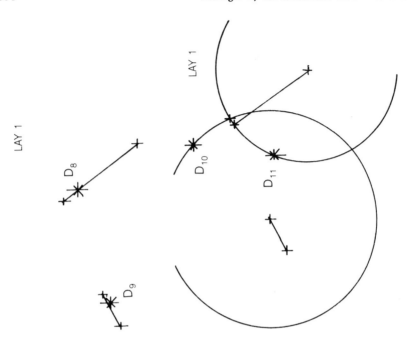

LAY 1

LAY 1

D_8

D_9

D_{10}

D_{11}

FIGURE 5.35g

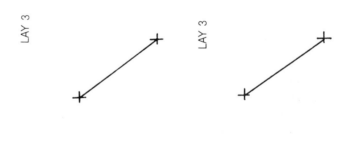

LAY 3

LAY 3

D_7

FIGURE 5.35f

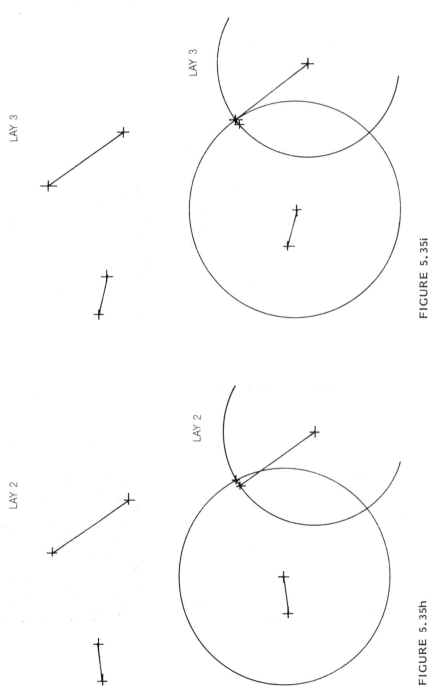

FIGURE 5.35i

FIGURE 5.35h

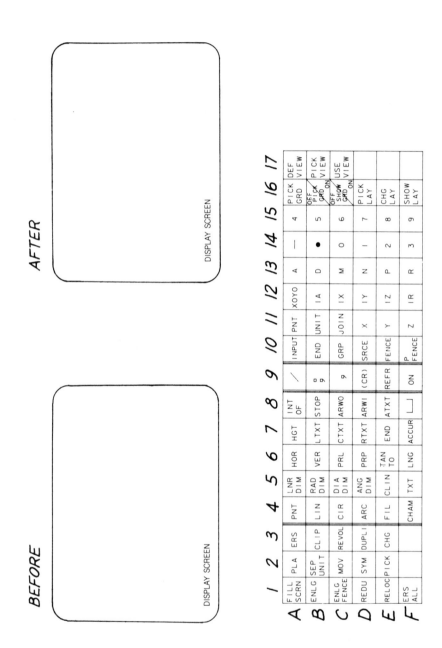

FILL SCRN	PLA	ERS	PNT	LNR DIM	HOR	HGT	INT OF	/	INPUT PNT	XOYO	A	—	4	PICK GRD	DEF VIEW
ENLG	SEP UNIT	CLIP	LIN	RAD DIM	VER	LTXT	STOP	⌐9	END UNIT	IA	D	●	5	OFF PICK GRD ON	PICK VIEW
ENLG FENCE	MOV	REVOL	CIR	DIA DIM	PRL	CTXT	ARWO	9	GRP JOIN	IX	M	O	6	OFF SHOW GRD ON	USE VIEW
REDU	SYM	DUPLI	ARC	ANG DIM	PRP	RTXT	ARWI	(CR)	SRCE X	IY	N		7	PICK LAY	
RELOC	PICK	CHG	FIL	CLIN	TAN TO	END	ATXT	REFR	FENCE Y	IZ	P	2	8	CHG LAY	
ERS ALL			CHAM	TXT	LNG		ACCUR ⌐	ON	P FENCE Z	IR	R	3	9	SHOW LAY	

FIGURE 5.36

5.9 VELOCITY ANALYSIS USING INTERACTIVE COMPUTER GRAPHICS

Since the relative-velocity method of finding velocities of mechanisms is the easiest to use, it will also be used on the ICGS. For instance, if the linear velocity of point A on link 2 in Fig. 5.32a is 10 in./sec CCW and the output linear velocity of point B on link 4 is to be determined using an ICGS, the following procedure could be followed.

Again, link 2 would be rotated incrementally (the number of increments would determine the amount of accuracy), in this case, every 30° or 12 rotated increments. The same kinematic motion analysis outlined previously would be performed first. Since the direction and magnitude of $V_{B4/1}$ are known, and the directions of $V_{B4/B2}$ and $V_{B4/1}$ are known, it would be necessary to place the linear velocity of point A on link 2 ($V_{A2/1}$) perpendicular to link 2 with a scale factor. In this case, the scale factor is 1 in. = 10 in./sec; therefore, a 1-in. line is placed at point $O_{2/1}$ in the direction of the angular velocity. Point $O_{2/1}$ will be used as the pole (representing a pivot point between the frame or link 1 and the other connecting links of the mechanism). The PLACE LINE PERPENDICULAR LONG 1 / UNIT D_1 END D_2 D_3 [CR] command is used (Fig. 5.37a). Using the same command, a similar line is placed on each of the other 11 layers. The results are shown in Fig. 5.37b.

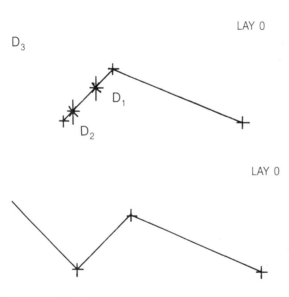

FIGURE 5.37a

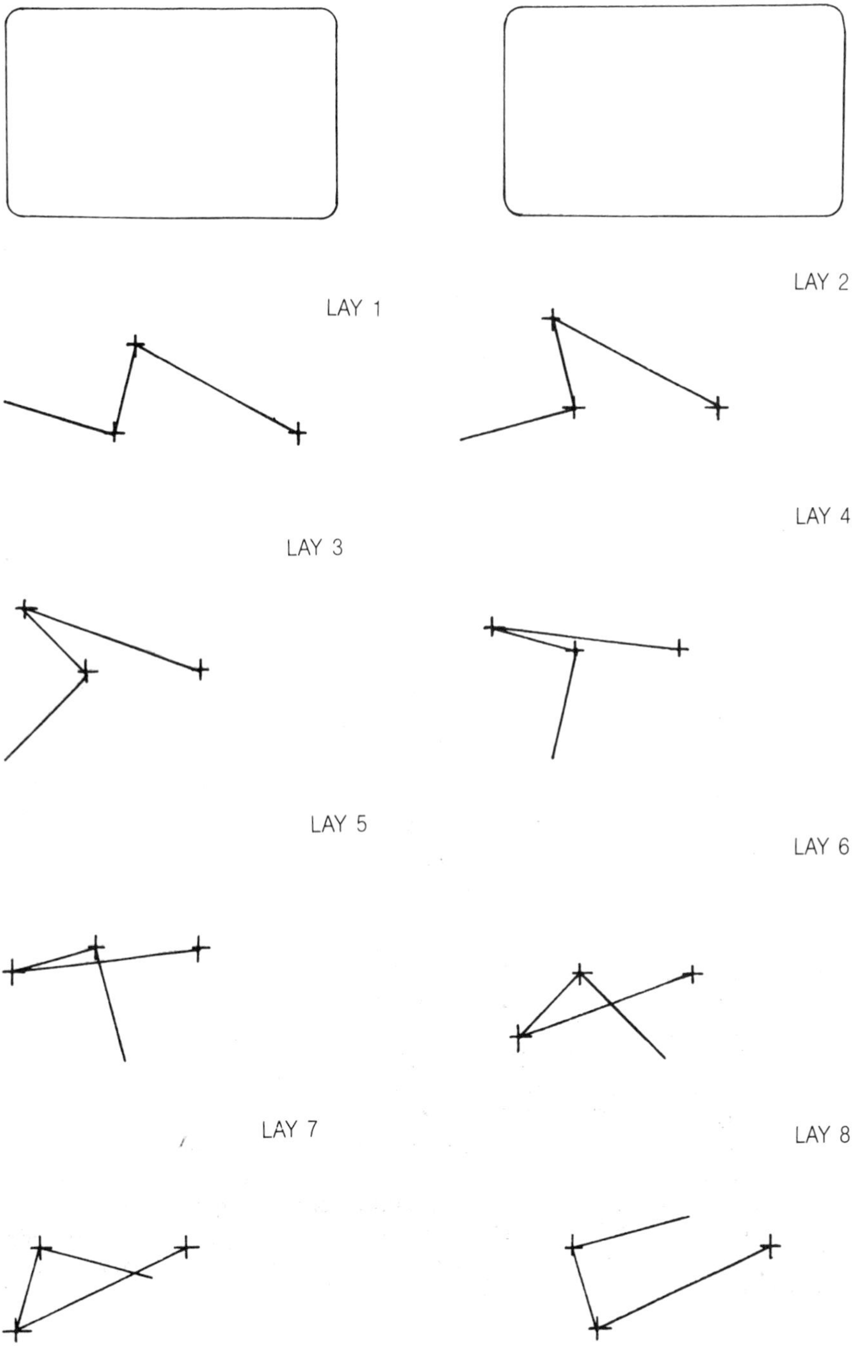

LAY 1

LAY 2

LAY 3

LAY 4

LAY 5

LAY 6

LAY 7

LAY 8

FIGURE 5.37b

LAY 9

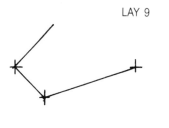

LAY 10

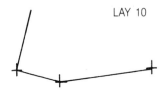

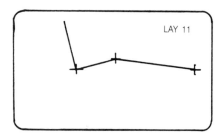

LAY 11

FIGURE 5.37b   (Continued)

The path of point B on link 4 traces a horizontal line while the direction of $V_{B4/A2}$ is constantly changing, but it is always perpendicular to link 3. Therefore, the same procedure can be followed for placing $V_{B4/A2}$ perpendicular to link 3 as was done when $V_{A2/1}$ was placed perpendicular to link 2, previously. The command PLACE LINE PERPENDICULAR / UNIT D_1 END D_2 D_3 [CR] can be used for determining the direction of $V_{B4/A2}$ since its magnitude is unknown. Its endpoint would be the endpoint of the $V_{A2/1}$ vector, as shown in Fig. 5.38a.

This procedure is performed on each of the remaining 11 layers, resulting in Fig. 5.38b. Now a horizontal line of any length is placed at point $O_{2/1}$. This horizontal line represents the path of $V_{B4/1}$, since its direction is known but not its magnitude. The PLANE LINE HORIZONTAL / UNIT END D_1 INPUT D_2 [CR] command is used to perform this task and this procedure must be repeated on each of the remaining layers (Figs. 5.39a and 5.39b).

Since each layer on the ICGS represents an incremental angular displacement of 30° of the input link of the mechanism, its corresponding velocity polygon is also constructed on each layer. To obtain the magnitudes of the linear velocities $V_{B4/A4}$ and $V_{B4/1}$, one must clip the two intersecting lines of each velocity polygon constructed on each layer by using the CLIP LINE INTERSECTION OF / UNIT D_1 D_2 [CR]

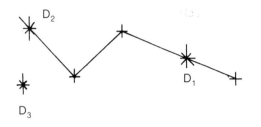

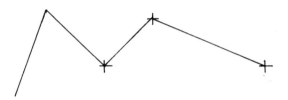

FIGURE 5.38a

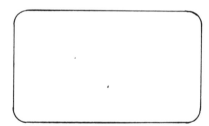

LAY 1

LAY 2

LAY 3

LAY 4

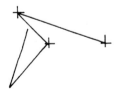

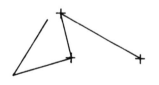

FIGURE 5.38b

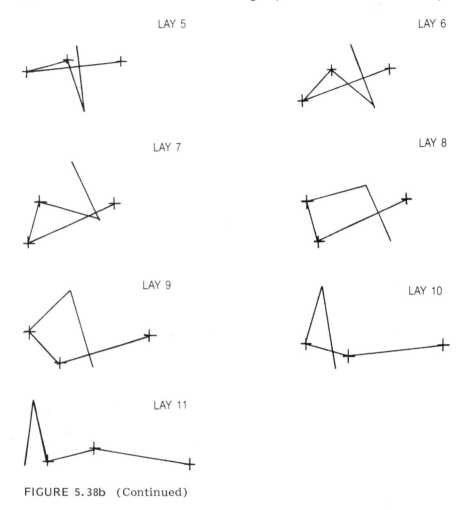

FIGURE 5.38b (Continued)

command (Fig. 5.39c). Figure 5.39b shows the results. Now it is a matter of measuring the line representing $V_{B4/1}$ on the polygon to obtain the magnitude of $V_{B4/1}$. This is done by using the MEASURE LINE / UNIT D_1 [CR] command (Fig. 5.39d) and multiplying this measured length by the scale factor (1 in. = 10 in./sec) to obtain $V_{B4/1}$ for each incremental rotation of the mechanism.

The scale factor could be a part of the MEASURE LINE command, which would give the measurement in units of velocity instead of length. To do this, a calculate expression command would be written using the particular ICGS's software language and this expression would be a part of the MEASURE LINE command.

LAY 0

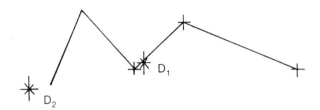

LAY 0

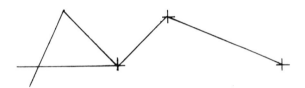

FIGURE 5.39a

LAY 1

LAY 2

LAY 3

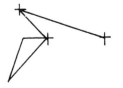

LAY' 4

FIGURE 5.39b

LAY 5   LAY 6

LAY 7   LAY 8

   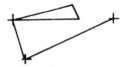

LAY 9

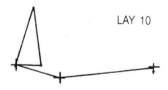   LAY 10

LAY 11

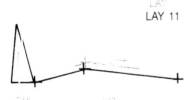

FIGURE 5.39b (Continued)

LAY 0

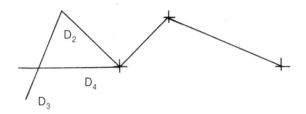

LAY 0

FIGURE 5.39c

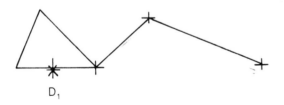

FIGURE 5.39d

5.10 ACCELERATION ANALYSIS USING
INTERACTIVE COMPUTER GRAPHICS

An ICGS can be used to determine the acceleration of the various links that make up a mechanism in a similar manner. For example, the four-bar linkage in Fig. 5.32a will be used again. Link 2 has an angular velocity of 10 in./sec in the counterclockwise direction and an angular acceleration of 20 in./sec per second counterclockwise.

The normal acceleration ($a^n = r\omega^2 = V^2/r$) and tangential acceleration ($a^t = r\alpha$) are vectors which are perpendicular to each other; the direction of the normal acceleration's vector is always along the given link toward the pivot point between the given link and the link representing the frame, while the direction of the tangential acceleration is always perpendicular to the given link.

Thus $a^n_{A2/1} = 1.25(10)^2 = 125$ in./sec^2 and this vector is placed as a 1.25-in. line along link 2 using a scale factor of 1 in. = 100 in./sec^2. (Fig. 5.40a). Similarly, $a^t_{A2/1} = 1.25(20) = 25$ in./sec^2 and this vector is placed as a 0.25-in. line perpendicular to link 2 (Fig. 5.40b). Again, point $O_{2/1}$ is used as the pole for this acceleration polygon and the PLACE LINE PARALLEL LONG 1.25 / UNIT D_1 INPUT END D_2 D_3 [CR] and PLACE LINE PERPENDICULAR LONG 0.25 / UNIT D_4 INPUT END D_5 D_6 [CR] commands are used.

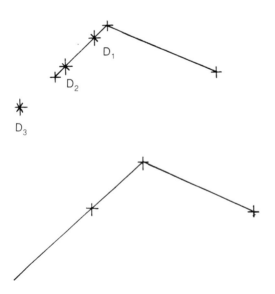

FIGURE 5.40a

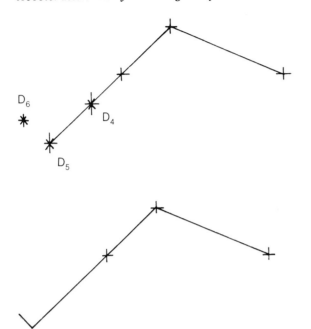

FIGURE 5.40b

Similarly, $a^n_{B4/A2} = V^2_{B4/A2} \div r_{B4/A2} = 46.26$ in./sec^2, and this value can be placed at the endpoint of the previous line representing $a^t_{A2/1}$ by using the PLACE LINE PARALLEL LONG 0.46 / UNIT D_1 INPUT END $D_2 D_3$ [CR] command and is shown in Fig. 5.41a. The magnitude of $a^t_{B4/A2}$ is unknown because $\alpha_{3/1}$ is not given or solvable, but its direction is always perpendicular to link 3. Therefore, a perpendicular line of any length is drawn from the endpoint of $a^n_{B4/A2}$ in the direction perpendicular to link 3. The PLACE LINE PERPENDICULAR / UNIT D_1 INPUT END $D_2 D_3$ [CR] command is used (Fig. 5.41b).

To closeup this acceleration polygon, the path of the linear acceleration of point B on link 4 ($a_{B4/1}$) is drawn. Figure 5.42 shows this path as a horizontal line intersecting the line representing $a^t_{B4/A2}$. The PLACE LINE HORIZONTAL / INPUT END $D_1 D_2$ [CR] command is used to draw this horizontal line.

The two intersecting lines must be clipped to form the completed acceleration polygon of this four-bar linkage for this particular posi-

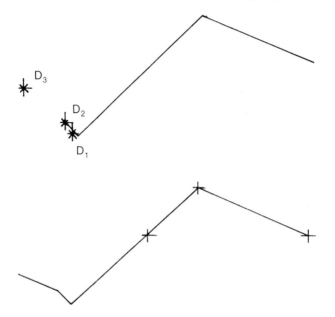

FIGURE 5.41a

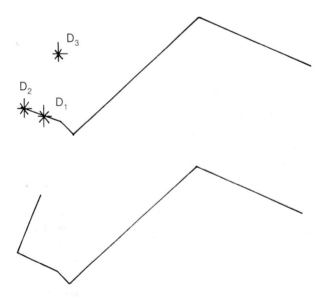

FIGURE 5.41b

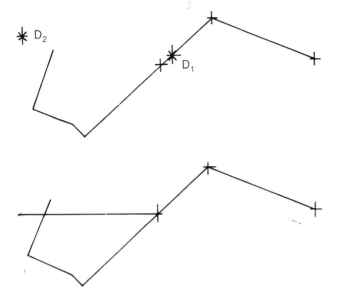

FIGURE 5.42

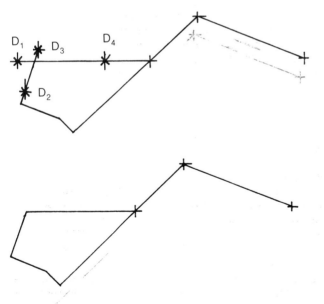

FIGURE 5.43

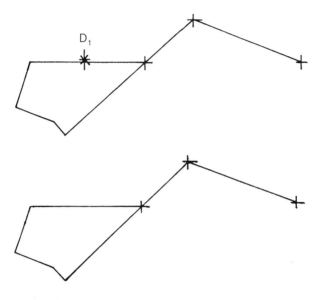

FIGURE 5.44

tion in time. The CLIP LINE INTERSECTION OF / UNIT D_1 D_2 [CR] and CLIP LINE INTERSECTION OF / UNIT D_3 D_4 [CR] commands are used to clip and complete the acceleration polygon (Fig. 5.43).

The acceleration of point B on link 4 is obtained by measuring the line representing $a_{B4/1}$ in this completed acceleration polygon. The MEASURE LINE / UNIT D_1 [CR] command is used with the scale factor of 1 in. = 100 in./sec^2 to give the magnitude of $a_{B4/1}$ (Fig. 5.44).

5.11 KINETIC (FORCE) ANALYSIS USING INTERACTIVE COMPUTER GRAPHICS

It is required to determine the inertia forces for each of the links of the mechanism shown in Fig. 5.45 at the given position using an ICGS. The resulting acceleration polygon obtained by using the procedure mentioned in the previous sections is shown on the ICGS's display screen in Fig. 5.46a. The acceleration images of the links are shown in the acceleration polygon when link 2 drives the mechanism at 2300 rpm. The centers of gravity g_2, g_3, and g_4 are located on the acceleration images of links 2, 3, and 4 such that the accelerations of the mass centers may be determined. The orientation of the cartesian coordinate system in Fig. 5.45 permits the location of the centers of

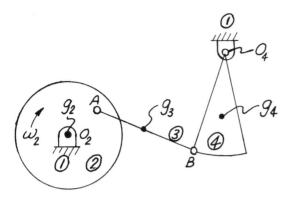

FIGURE 5.45

gravity g_3 and 6_4 exactly on links 3 and 4, respectively, by using the
PLACE POINT / INPUT X Y [CR] command, as shown in Fig. 5.46b.
Links 3 and 4 are reduced proportionally to fit into the acceleration
polygon by a ratio of the length of AB measured on the acceleration
polygon divided by the length of AB on the kinematic drawing, or
$1.18/1.45 = 0.814$. The REDUCE R.814 / UNIT D_1 [CR] and REDUCE
R.814 /UNIT D_2 D_3 D_4 [CR] command are used to reduce links 3 and
4 and the results are on the display screen of the ICGS in Fig. 5.47.
Then links 3 and 4 are placed on the acceleration polygon by using
the commands MOVE / UNIT D_1 END D_2 D_3 INPUT END D_4 D_5 [CR]

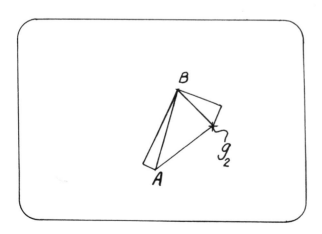

FIGURE 5.46a

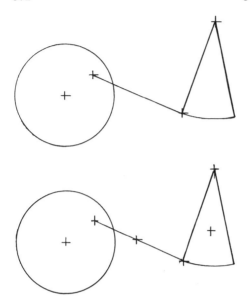

FIGURE 5.46b

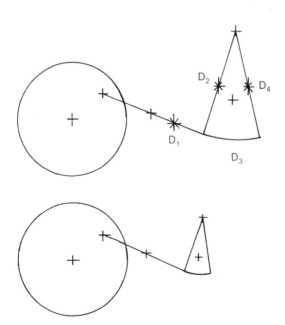

FIGURE 5.47

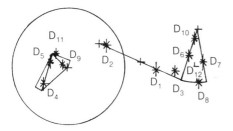

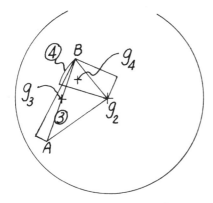

FIGURE 5.48

and MOVE / UNIT D_6 D_7 D_8 END D_9 D_{10} INPUT END D_{11} D_{12} [CR], respectively (Fig. 5.48).

The magnitudes of a_{g2}, a_{g3}, and a_{g4} are measured from the acceleration polygon by drawing a line to the point representing the center of gravity of the particular link and measuring this resulting line. The PLACE LINE / INPUT END D_1 POINT D_2 [CR] and MEASURE LINE / UNIT D_5 [CR] commands are used to perform this task (Fig. 5.49). The angular accelerations $\alpha_{2/1}$, $\alpha_{3/1}$, and $\alpha_{4/1}$ and the magnitudes of the inertia forces $F_{O2/1}$, $F_{O3/1}$, and $F_{O4/1}$ are calculated and the results are summarized:

$$\underline{a}_{g2/1} = b$$

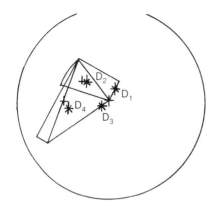

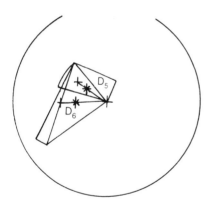

FIGURE 5.49

$$\underline{a}_{g_{3/1}} = 9500 \text{ ft/sec}^2$$

$$\underline{a}_{g_{4/1}} = 6700 \text{ ft/sec}^2$$

$$\alpha_{2/1} = 0$$

$$\alpha_{3/1} = \underline{a}_{B_{4/}}_{A_2} /AB = 15,700/\frac{8}{12} = 23,600 \ \frac{\text{rad}}{\text{sec}^2} \ (\text{CCW})$$

$$\alpha_{4/1} = a_{B_{4/1}} / O_4 B = 9000 / \frac{8}{12} = 13,500 \ \frac{rad}{sec^2} \ (CW)$$

$$F_{O_{2/1}} = m_2 a_{g_{2/1}} = \frac{W_2}{g} a_{g_{2/1}} = 0$$

$$F_{O_{3/1}} = m_3 a_{g_{3/1}} = \frac{W_3}{g} a_{g_{3/1}} = \frac{4}{32.2} \times 9500 = 1180 \ lb$$

$$F_{O_{4/1}} = m_4 a_{g_{4/1}} = \frac{W_4}{g} a_{g_{4/1}} = \frac{8}{32.2} \times 6700 = 1670 \ lb$$

The locations from the mass center of the lines of action of the inertia forces are determined by e_3 and e_4 from

$$e_3 = \frac{I_{3/1} \alpha_{3/1}}{F_{O_{2/1}}} = \frac{0.006(23,600)}{1180} = 0.12 \ ft = 1.44 \ in.$$

$$e_4 = \frac{I_{4/1} \alpha_{4/1}}{F_{O_{4/1}}} = \frac{0.026(13,500)}{1670} = 0.21 \ ft = 2.52 \ in.$$

Since for link 2, both $a_{g2/1}$ and $\alpha_{2/1}$ are zero, there is no inertia force or inertia couple. In Fig. 5.50, $F_{O3/1}$ and $F_{O4/1}$ are shown on

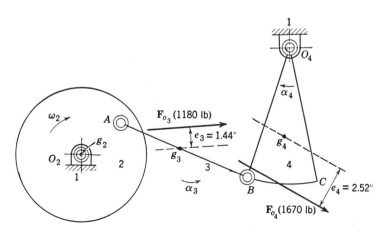

FIGURE 5.50

links 3 and 4 opposite in sense to $a_{g3/1}$ and $a_{g4/1}$; $e_{3/1}$ and $e_{4/1}$ are oriented as shown in order that the senses of the moments $F_{O3/1}e_{3/1}$ and $F_{O4/1}e_{4/1}$ are opposite, respectively, to the senses of $\alpha_{3/1}$ and $\alpha_{4/1}$. To place $F_{O3/1}$ and $F_{O4/1}$ on the kinematic drawing appearing on the display screen of the ICGS, the following commands are used:

> PLACE LINE PARALLEL 1.44 LONG 0.85 / UNIT D_1 POINT D_2
> INPUT D_3 [CR]
> PLACE LINE PARALLEL 2.52 LONG 1.1 / UNIT D_4 POINT D_5 INPUT
> D_6 [CR]

The results are shown in Fig. 5.51.

The determination of inertia forces is only a preliminary step in the determination of the forces producing the accelerationed motion indicated by the acceleration polygon. For example, the force $F_{O4/1}$ is the equilibrant of forces acting on link 4 through the connections at B and $O_{4/1}$. Similarly, the inertia force $F_{O3/1}$ is the equilibrant of

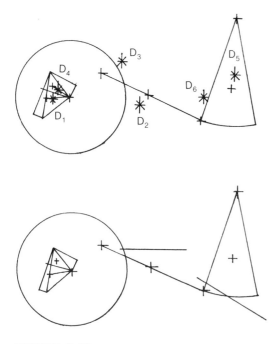

FIGURE 5.51

forces acting on link 3 through A and B. The three forces on link 3 produce tension and bending on the link. If the speed of the driving link 2 is increased, the inertia force and the forces at the connections also increase, with the possibility of exceeding the strength of the link in tension and bending. The determination of the unknown forces at the connections from the known inertia force will now be done.

In the force analysis of a complete mechanism, the free-body diagram of each link should be shown to indicate the forces acting on the link. For each link, a polygon of free force vectors may be drawn and should close for static equilibrium. The construction of the polygon on an ICGS permits the determination of the unknown forces.

The principle of superposition is used in determining the unknown forces on each link. Figure 5.52 shows the graphical determination of forces for the case in which only inertia force $F_{O_{4/1}}$ acts on the mechanism.

From Fig. 5.52 a free-body diagram of link 4 (Fig. 5.53a, 5.53b, and 5.53c) is drawn first, since it include the force $F_{O_{4/1}}$. Link 4 is acted upon by three forces, two of which are the unknown forces of links 1 and 3 acting on link 4 through the bearings at $O_{4/1}$ and B. $F'_{1/4}$ is the force of link 1 on link 4 (the subscript 1/4 designates "link 1 acting on link 4"), and $F'_{3/4}$ is the force of link 3 on link 4. The line of action of $F'_{3/4}$ is known since the free-body diagram of link 3 shows link 3 to be acted on by two forces that must lie along line AB. The three forces of link 4 are concurrent at the intersection of the known lines of action of $F_{O_{4/1}}$ and $F'_{3/4}$. Thus the intersection of these direction lines determines the magnitudes of the unknown forces, and the senses are determined from the condition that for static equilibrium the resultant is zero when the force vectors are added.

Since link 3 is acted on by two forces in static equilibrium, $F'_{2/3}$ is equal and collinear but opposite in sense to $F'_{4/3}$. The magnitudes of these unknowns are determined from $F'_{3/4}$, which is equal but opposite in sense to $F'_{4/3}$ by the law that acton and reaction are equal but opposite. The PLACE LINE PARALLEL LONG 0.9 / UNIT POINT $D_1 D_2 D_3$ [CR] command is used to place the line representing the force $F'_{3/4}$ to point B on link 3. The same command is used to place this line to point A on link 3 and this line will represent $F'_{2/3}$ (Figs. 5.54a and 5.54b).

The free-body diagram of link 2 shows that two forces, at $O_{2/1}$ and A, act on the link as well as the shaft torque T'_s. $F'_{3/2}$ is known from $F'_{2/3}$. Since only two forces act on link 2, $F'_{1/2}$ is equal and opposite to $F'_{3/2}$ for static equilibrium of forces; these forces are not collinear and therefore act as a clockwise couple. For static equilibrium of torques, T'_s is a counterclockwise couple equal and opposite to $F'_{3/2}d'$. Again, the line representing $F_{3/2}$ is moved and duplicated at point A on link 3 by using the PLACE LINE PARALLEL LONG 0.9 /

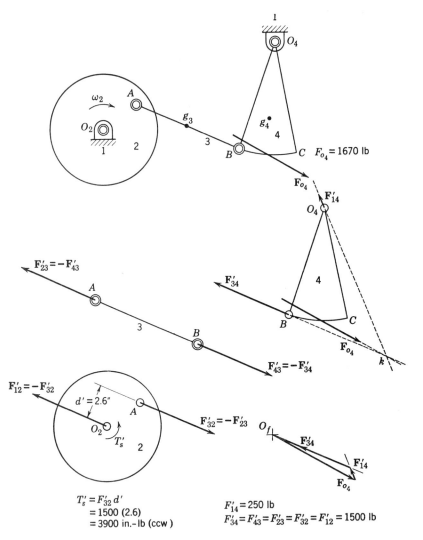

$T'_s = F'_{32} d'$
  $= 1500 (2.6)$
  $= 3900$ in.-lb (ccw)

$F'_{14} = 250$ lb
$F'_{34} = F'_{43} = F'_{23} = F'_{32} = F'_{12} = 1500$ lb

FIGURE 5.52

UNIT POINT D_1 D_2 [CR] command and a similar line is drawn at point $O_{2/1}$ by using the same command (Figs. 5.55a and 5.55b).

   In the force analysis due to $F_{O3/1}$ as shown in Fig. 5.56, link 3 is acted on by three forces and link 4 by two forces. The free-body diagram of link 3 is drawn first, since $F_{O3/1}$ is known, and the polygon

LAY 0 ⌐⌐ 1 ⌐ 2

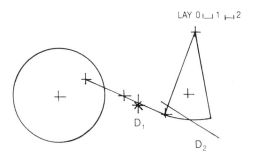

LAY 0 ⌐⌐ 1

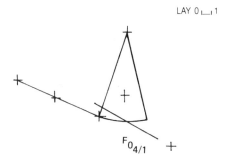

FIGURE 5.53a

shown is drawn to determine the unknown forces. On an ICGS, this analysis is performed on different layers from those used of the first analysis. This makes the resultant force analysis by superposition much easier.

The resultant effects due to $F_{O_{3/1}}$ and $F_{O_{4/1}}$ are shown on the ICGS's display screen in Fig. 5.57a, in which the separate solutions (Figs. 5.57b and 5.57c) are superposed. The force analysis of the mechanism with $F_{O_{4/1}}$ acting can be placed on the force analysis with $F_{O_{3/1}}$ acting by using the SHOW OVERLAY n [CR] command, since the force analysis with $F_{O_{4/1}}$ & $F_{O_{3/1}}$ acting was constructed on different overlays.

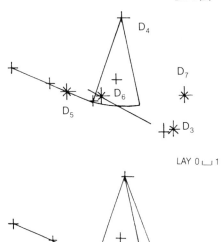

FIGURE 5.53b

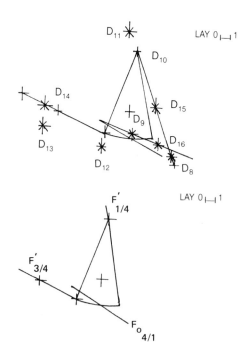

FIGURE 5.53c

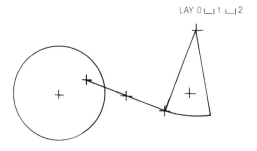

LAY 0 ⌞⌟ 1 ⌞⌟ 2

LAY 1

FIGURE 5.54a

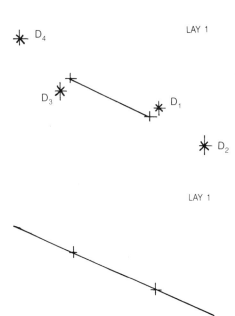

FIGURE 5.54b

LAY 0 ⌐1 ⌐2

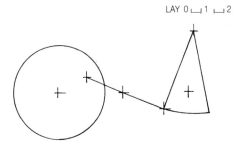

LAY 2

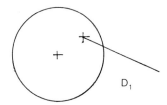

D_1

FIGURE 5.55a

LAY 1 ⌐2

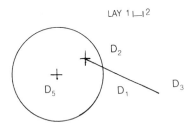

D_2

D_3

D_5 D_1

LAY 2

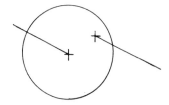

FIGURE 5.55b

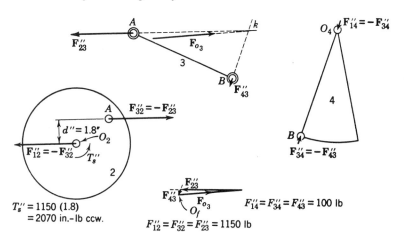

FIGURE 5.56

The PICK LAY command activates the particular overlay on which the geometry will be constructed. Other overlays can be activated and shown with the particular picked overlay, but the constructed geometry will appear only on the picked overlay. The PICK LAY n [CR] command is used to pick the construction overlay and the SHOW LAY n [CR] command shows overlay n_1 on top of overlay n_2 and both are visible to the viewer, as shown in Fig. 5.58a.

The MOVE LINE commands are used (Figs. 5.58b and 5.59b) to form the force polygons for links 1, 2, 3, and 4. From these force polygons the bearing reactions are determined from the resultant

LAY 0 ⌐ 3

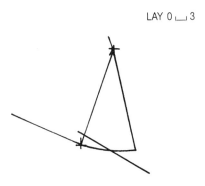

FIGURE 5.57a

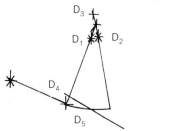

FIGURE 5.57b

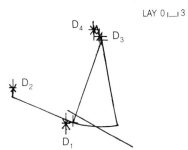

LAY 0

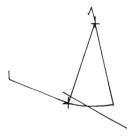

FIGURE 5.57c

384

LAY 1 ⌐⌐ 4

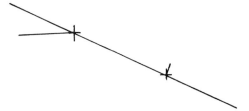

FIGURE 5.58a

LAY 1 ⌐⌐ 4

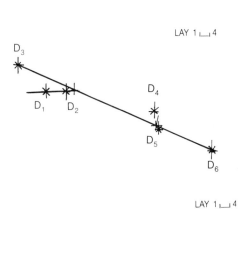

LAY 1 ⌐⌐ 4

FIGURE 5.58b

LAY 1 |___| 4

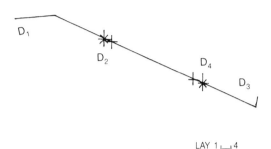

LAY 1 |___| 4

FIGURE 5.58c

free-body diagrams shown in Figs. 5.58c and 5.59c. At $O_{2/1}$, the
bearing force is 2600 lb; at A, 2600 lb; at B, 1500 lb; and at $O_{4/1}$,
180 lb. Each bearing should be selected or designed for the maximum
force as determined from force analysis of the linkage in its various
phases.

In a complete force investigation, the drive link is advanced by,
say, 15° intervals, and a force analysis is made for each corresponding
phase of the mechanism. A subroutine could be worked out that would
calculate the velocities, accelerations, and kinetic forces involved at
each point of the linkage for each corresponding phase of the mechanism directly from the ICGS's geometry.

For example, for the four-bar linkage shown in Fig. 5.60, one
could derive the piston velocity and acceleration analytically. Let r
be the crank length and nr be the connecting-rod length to crank
length. Suppose the crank to be at any angle θ with the line of stroke,
and φ the corresponding inclination of the connecting rod. The term
x is the distance from the center of the crosshead pin to the center of
the crankshaft. At midstroke, evidently, x = nr. At any crank angle

LAY 2 ⌐ 5

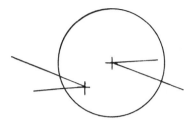

FIGURE 5.59a

LAY 2 ⌐ 5

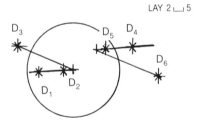

LAY 2 ⌐ 5

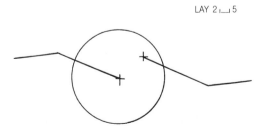

FIGURE 5.59b

LAY 2⌊⌋5

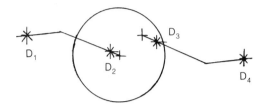

D_1 D_2 D_3 D_4

LAY 2⌊⌋5

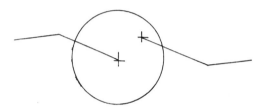

FIGURE 5.59c

θ the piston displacement s from midposition $= x - nr$. From Fig.
5.60, $x = r \cos \theta + nr \cos \phi$, and the piston displacement

$$s = x - nr$$
$$= r \cos \theta + nr \cos \phi - nr$$
$$= r (\cos \theta + n \cos \phi - n)$$

Also,

$$\sin \theta = \frac{h}{r} \quad \text{and} \quad \sin \phi = \frac{h}{nr}$$

Now

$$\cos \phi = \sqrt{1 - \sin^2 \phi} = 1 - \frac{\sin^2 \theta}{n^2} = \frac{1}{n} \sqrt{n^2 - \sin^2 \theta}$$

Substituting this value of $\cos \phi$ in $s = r (\cos \theta + n \cos \theta - n)$ yields
piston displacement $= s = r(\cos \theta + \sqrt{n^2 - \sin^2 \theta} - n)$. Thus the piston displacement is in terms of the crank angle.

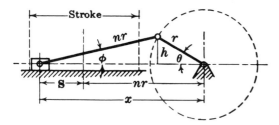

FIGURE 5.60

The piston velocity is equal to ds/dt, where s is the piston displacement. Substituting the value of s from $s = r(\cos\theta + \sqrt{n^2 - \sin^2\theta} - n)$ gives

$$\text{Piston velocity} = v = \frac{d}{dt}\left[r(\cos\theta + \sqrt{n^2 - \sin^2\theta} - n)\right]$$

$$v = r\left[-\sin\theta\,\frac{d\theta}{dt} + \frac{1}{2}(n^2 - \sin^2\theta)^{-1/2} \times \frac{d(n^2 - \sin^2\theta)}{dt}\right]$$

$$v = -r\,\frac{d\theta}{dt}\left(\sin\theta + \frac{2\sin\theta\cos\theta}{2\sqrt{n^2 - \sin^2\theta}}\right)$$

$$v = -r\omega\left(\sin\theta + \frac{\sin 2\theta}{2\sqrt{n^2 - \sin^2\theta}}\right)$$

since $d\theta/dt = \omega$ = angular velocity of the crank. An approximate form of this equation is obtained by neglecting $\sin^2\theta$ in the denominator. The error involved is not very large, the value of n in engine design seldom being less than 4, and $\sin^2\theta$ being equal to 1 as a maximum. The equation

$$v = -r\omega\left(\sin\theta + \frac{\sin 2\theta}{2\sqrt{n^2 - \sin^2\theta}}\right)$$

reduces to the following form:

$$\text{Piston velocity} = v = -r\omega\left(\sin\theta + \frac{\sin 2\theta}{2n}\right)$$

The piston acceleration is equal to dv/dt. By taking the exact velocity equation,

$$v = -r\omega \left(\sin\,\theta + \frac{\sin\,2\theta}{2\sqrt{n^2 - \sin^2\theta}} \right)$$

differentiating, and dividing by dt, it can be shown that

$$\text{Piston acceleration} = a = r\omega^2 \left[\cos\,\theta + \frac{\cos^4\theta + \cos\,2\theta(n^2 - 1)}{(n^2 - \sin^2\theta)^{3/2}} \right]$$

Treating the approximate equation

$$v = -r\omega \left(\sin\,\theta + \frac{\sin 2\theta}{2n} \right)$$

in the same manner, then

$$a = \frac{d}{dt} \left[r\omega (\sin\,\theta + \frac{\sin\,2\theta}{2n}) \right]$$

$$= r\omega^2 \left(\cos\,\theta + \frac{\cos\,2\theta}{n} \right)$$

which is so much simpler that it is generally used where extreme accuracy is not required.

The following subroutine is written using BASIC as the computer programming language. This subroutine contains the previously derived velocity and acceleration equations and this subroutine can be directly incorporated with the geometry constructed on the ICGS.

This velocity and acceleration subroutine would be directly connected to the input and connecting links of this constructed four-bar linkage appearing on the display screen of the ICGS. For instance, the first line digitized would measure the length r of the link representing the input link, while the second line digitized would measure the length of nr or 1 of the link representing the connecting link. The MEASURE LINE command would be used to measure the input and connecting links' lengths and these values would be placed as part of the data base of the subroutine (in this case, r = 2 in. and 1 = 4 in.). The angular velocity ω (in this case, 10 rad/sec) must be inputted to complete the data base and begin the velocity and acceleration calculations. The rotational increment is preset as 0.1 rad of the input link for this particular subroutine.

The resultant velocity and acceleration for point B on the ouput link (link 4) for each 0.1 rad of rotation is listed on the display screen of the ICGS as shown below. Another subroutine could be written to calculate the inertia forces present at the pivot points of this linkage and would use the calculated accelerations from the previous subroutine as its data base for calculating these inertia forces.

```
WHAT'S R
?2
WHAT'S L
?4
WHAT'S OMEGA
?10
```

0.1	2.990015	297.0075
0.2	5.920478	288.1194
0.3	8.733616	273.6009
0.4	11.37515	253.8829
0.5	13.79587	229.5467
0.6	15.95304	201.3029
0.7	17.8116	169.9652
0.8	19.34499	136.4214
0.9	20.53578	101.6018
1	21.37591	66.44578
1.1	21.86663	31.86911
1.2	22.0181	-1.267818
1.3	21.84867	-32.18911
1.4	21.38394	-60.2288
1.5	20.6555	-84.85181
1.6	19.6996	-105.6694
1.7	18.55559	-122.4487
1.8	17.26435	-135.1163
1.9	15.86671	-143.7547
2	14.40194	-148.5937
2.1	12.90631	-149.9953
2.2	11.41192	-148.4335
2.3	9.945649	-144.4705
2.4	8.528441	-138.7288
2.5	7.174822	-131.8625
2.6	5.892754	-124.5261
2.7	4.683775	-117.3451
2.8	3.54343	-110.8879
2.9	2.461976	-105.6397
3	1.425323	-101.9815
3.1	.4161666	-100.1728
3.2	-.5847367	-100.3405
3.3	-1.597207	-102.4727
3.4	-2.640255	-106.4199
3.5	-3.730731	-111.9011
3.6	-4.882069	-118.5166

3.7	-6.103182	-125.7653
3.8	-7.397559	-133.0676
3.9	-8.762606	-139.7909
4	-10.18926	-145.2787
4.1	-11.66189	-148.8803
4.2	-13.15852	-149.981
4.3	-14.65133	-148.0318
4.4	-16.10746	-142.5759
4.5	-17.49001	-133.2722
4.6	-18.75937	-119.9149
4.7	-19.87459	-102.4471
4.8	-20.79493	-80.969
4.9	-21.48145	-55.74016
5	-21.89859	-27.17473
5.1	-22.01567	4.168971
5.2	-21.80823	37.60488
5.3	-21.25923	72.34103
5.4	-20.35997	107.5056
5.5	-19.11076	142.1765
5.6	-17.52122	175.4136
5.7	-15.61036	206.2916
5.8	-13.40619	233.9329
5.9	-10.94516	257.5389
6	-8.271176	276.4194
6.1	-5.434398	290.017
6.2	-2.489811	297.9276

The first column of values represents the position of link 2 (input link), while the second column of values represents the resulting linear velocities of point B on link 4 (output link) with respect to the particular positions of link 2. The third column of values represents the corresponding accelerations of point B on link 4 with respect to the particular positions of link 2. The minus sign signifies that the velocity and acceleration vectors are in the opposite direction of the displacement vector of link 4. In other words, link 4 is slowing down (decreasing in velocity) and/or decelerating (decreasing acceleration).

Bibliography

Barr, Paul. *CAD: Principles and Applications*. Prentice-Hall, Englewood Cliffs, N.J., 1985

Bowman, Daniel J. *The CAD/CAM Primer*. Howard W. Sams & Co., Inc., Indianapolis, Ind., 1983.

Besant, C. B. *Computer-Aided Design and Manufacture*. 2nd ed. Halsted Press, New York, 1983.

Bertoline, Gary R. *Fundamentals of CAD*. Delmar, Albany, N.Y., 1985.

Chasen, Sylvan H. *Geometric Principles and Procedures for Computer Graphic Applications*. Prentice-Hall, Englewood Cliffs, N.J., 1978.

Demel, John T. and Michael J. Miller. *Introduction to Computer Graphics*. Brooks-Cole, Monterey, Ca., 1984.

Foley, J. and A. Van Dam. *Fundamentals of Interactive Computer Graphics*. Addison-Wesley, Boston, 1982.

Gardan, Y. and M. Lucas. *Interactive Graphics in CAD*. Unpublished, 1984.

Giloi, Wolfgang K. *Interactive Computer Graphics: Data Structures, Algorithms, Languages*. Prentice-Hall, Englewood Cliffs, N.J., 1978.

Groover, M. and E. Zimmers. *CAD/CAM: Computer-Aided Design and Manufacturing*. Prentice-Hall, Englewood Cliffs, N.J., 1984.

House, William C. *Interactive Computer Graphics Systems*. Petrocelli, Princeton, N.J., 1982.

Iliffe, J. *Advanced Computer Design*. Prentice-Hall, Englewood Cliffs, N.J., 1983.

Krouse, John K. *Computer-Aided Design and Computer-Aided Manufacturing: The CAD/CAM Revolution*. Marcel Dekker, New York, 1982.

Lange, Jerome C. *Design Dimensioning with Computer Graphics Applications*. Marcel Dekker, New York, 1984.

Lange, Jerome C. and Dennis P. Shanahan. *Interactive Computer Graphics Applied to Mechanical Drafting and Design.* John Wiley and Sons, New York, 1984.

Pao, Y. C. *Elements of Computer-Aided Design and Manufacturing.* John Wiley and Sons, New York, 1984.

Preparata, Franco. *Fundamentals of Computer Engineering.* Harper & Row, New York, 1984.

Ryan, Daniel. *Computer-Aided Graphics and Design.* 2nd ed. Marcel Dekker, New York, 1985.

Scott, Joan E. *Introduction to Interactive Computer Graphics.* Wiley (Interscience), New York, 1982.

Sangiovanni-Vincentelli, Alberto. *Advances in Computer-Aided Engineering Design*, Vol. 1. Jai Press, Greenwich, Conn., 1983.

Tercholz, E. *CAD–CAM Handbook.* McGraw-Hill, New York, 1984.

Walker, B. S. *Interactive Computer Graphics.* Crane-Russak Co., New York, 1976.

Index

A

Acceleration, 321
Acceleration-image method, 328
Analysis, 6
Analytic geometry, 143
Angular acceleration, 321
Angular velocity, 321
Archimedes, 144
Area, 145
Area moments of inertia, 149

B

B-spline, 124
Blank space, 28
Boundary conditions, 237

C

Carriage return, 28
Cathode-ray tube, 9, 17, 19, 20
Center of gravity, 147
Center of mass, 147
Centroid, 147
Central processing unit, 7, 20
Centripetal acceleration, 322
Colon, 28

Comma

Comma, 28
Complex joint, 317
Compressive stress, 221
Computer, 6, 7
Connecting-link method, 327
Coriolis law, 329
Cursor, 9, 13

D

d'Alembert's principle, 333
Deceleration, 321
Design, 1
Designer, 1, 2
Digitizer, 12, 13
Digitizing pen, 4
Disk unit, 8, 20
Displacement, 321
Direct method, 327

E

Electronic pen, 17, 19
Electrostatic plotter, 12
Element, 259
Engineering, 1
Energy balance approach, 261
Epsilon, 222

F

Finite elements, 264
Finite element mesh, 264
Finite element method, 258
Finite element model, 6
First moment of inertia, 148
Fixed joint, 317
Flatbed plotter, 11
Floating-point unit, 20

G

Geometry, 143
Graphics level, 23
Graphics process/unit, 20
Graphics tablet, 17

H

Hard-copy printer, 17
Hidden edges, 129
Hooke's law, 224

I

Image control unit, 17
Impact plotter, 12
Ink jet plotter, 12
Instant center, 323
Interactive computer graphics,
   2, 4

K

Kennedy's theorem, 323
Keyboard, 8, 9, 17
Kinematics, 315
Kinetics, 315, 332

L

Laser plotter, 12
Light pen, 9
Linear acceleration, 321
Linear velocity, 321

M

Mass moments of inertia, 158
Maximum stress, 242
Mechanical design, 2
Mechanics, 143, 315
Menu, 9, 17, 19
Mesh, 128
Microprocessors, 7
Modes, 23
Modulus of elasticity, 224

N

NASTRAN, 264, 298
Neutral surface, 235
Nodal points, 260
Nodes, 260
Normal acceleration, 322
Normal stress, 221
Normal strain, 222

O

Operating system, 23
Origin, 28

P

Parallel-axis theorem, 155
Patches, 128
Photoplotters, 12
Pin joint, 317
Pixels, 10, 11
Plotters, 11
Poisson's ratio, 224
Polar moment of inertia for areas,
   151, 152
Postprocessing, 11
Principal planes, 250
Principal stresses, 250
Printer, 8, 19
Process of summation, 146
Product of inertia, 155
Products of mass inertia,
   164

R

Radius of gyration areas, 152
  mass moment of inertia, 160
  about a polar axis, 153
Raster-scan, 10, 11
Relative-velocity method, 328
Rectangular moments of iner-
    tia, 152, 153
Resolution method, 328

S

Section modulus, 236
Shear flow, 232·
Shear modulus, 225
Shear strain, 224
Shear stress, 221
Sigma, 221
Sliders, 317
Sliding plotters, 11
Simple joint, 317
Software, 13, 23
Special-function keyboard, 9
Statics, 315
Stick figure, 95
Stiff joint, 317
Storage tube, 10
Stress, 221
STRUDL, 298
Stylus, 9, 13
SUPERB, 299
System level, 23

T

Tablet, 9
Tabulated cylinders, 116
Tabular method, 325
Tangential acceleration, 322
Tape unit, 7, 8, 20
Tau, 221
Tensile stress, 221
Tolerance stack-up, 78
Transfer of axes:
  area moment of inertia, 156
  mass moments of inertia, 161

U

Uniform axial stress, 226
Uniform shear stress, 228

V

Variational approach, 260
Vector-refresh, 10
Velocity-image method, 328
View, 28, 30
Volume, 157

W

Weighted residuals approach,
    261
Wireframe, 95
Workstation, 8, 14, 15